MAKING SPACE

Space, Place, and Society
John Rennie Short, *Series Editor*

Other books in Space, Place, and Society

At Home: An Anthropology of Domestic Space
    Irene Cieraad, ed.

Becoming a Geographer
    Peter Gould

Constructions of "Nation": The Politics of Landscape in Singapore
    Lily Kong and Brenda S. A. Yeoh

Geographical Voices
    Peter Gould and Forrest R. Pitts, eds.

Geography Inside Out
    Richard Symanski

The Global Crisis in Foreign Aid
    Richard Grant and Jan Nijman, eds.

Imagined Country: Environment, Culture, and Society
    John Rennie Short

New Worlds, New Geographies
    John Rennie Short

Placing Autobiography in Geography
    Pamela Moss, ed.

Pleasure Zones: Bodies, Cities, Spaces
    David Bell, ed.

Putting Health into Place: Landscape, Identity, and Well-being
    Robin A. Kearns and Wilbert M. Gesler, eds.

*Second Nature: The History and Implications of
    Australia as Aboriginal Landscape*
    Lesley Head

MAKING SPACE

REVISIONING THE WORLD, 1475–1600

JOHN RENNIE SHORT

SYRACUSE UNIVERSITY PRESS

Syracuse University Press
Syracuse, New York 13244-5160

Copyright © 2004 by John Rennie Short
All Rights Reserved

First Edition 2004
04 05 06 07 08 09   6 5 4 3 2 1

The paper used in this publication meets the minimum requirements of American National Standard for Information Sciences—Permanence of Paper for Printed Library Materials, ANSI Z39.48–1984.∞™

Library of Congress Cataloging-in-Publication Data
Short, John R.
  Making space : revisioning the world, 1475–1600 / John Rennie Short.—1st ed.
    p. cm. — (Space, place, and society)
  Includes bibliographical references and index.
  ISBN 0–8156–3023–9 (cloth : alk. paper)
  1. Cartography—History—16th century. 2. Cartography—History—To 1500. 3. Cartography—Social aspects. I. Title. II. Series.
GA231 .S46 2004
912'.094'09031—dc22   2003024727

Manufactured in the United States of America

It is requisite to have a knowledge of both latitude and longitude.[1]

— CLAUDIUS PTOLEMY

Without the habit of conceptualizing space, a traveller going to war or work could not link his separate impressions to the nature of his route as a whole or extend them imaginatively to the unseen parts of the area through which he was passing; a man could not visualise the country to which he belonged; a landowner, unable to "see" his properties as a whole was not concerned to concentrate his scattered holdings by sale or exchanges; a ruler unable to "see" his kingdom was not perturbed by bargaining away provinces that map-conscious generations were able to see as essential to strategic frontiers; governments informed by verbal descriptions, were unable to judge the resources in men and materials of their rivals.[2]

— J. R. HALE

John Rennie Short is chair of the Department of Geography and Environmental Systems at the University of Maryland, Baltimore County. The recipient of many awards and fellowships, in 2002 Short was named to the Leverhulme Visiting Professorship to the UK. He is the author or editor of twenty-three books as well as many articles exploring geography and its complex intersections with human society, culture, environment, and history. Professor Short is also series editor for the Space, Place, and Society series published by Syracuse University Press. His most recent publications include *Global Dimensions: Space, Place and the Contemporary World*, published in 2001 by Reaktion; *Globalization and the Margins*, coedited with Richard Grant and published in 2002 by Palgrave; and *The World Through Maps*, published in 2003 by Firefly.

CONTENTS

Illustrations ix

Acknowledgments xv

1. Introduction 1
2. Coordinating the World 9
3. Encompassing the World 34
4. Mapping the World 68
5. Navigating the Seas 109
6. Surveying the Land 129
7. Annexing Territories 142
8. Conclusions 150

Appendix
A Selection of the Renaissance Translations of
Ptolemy's *Geography*, 1475–1600 159

Notes 161

Selected Bibliography 167

Index 175

ILLUSTRATIONS

1. Sebastiano del Piombo, *Cardinal Bandinello Sauli, His Secretary, and Two Geographers*, 1516   2
2. The Hereford Cathedral Mappa Mundi, ca. 1300   10
3. Ptolemy's world map projections   14
4. Latitude and longitude in Ptolemy's *Geography*; detail from manuscript fragment, 1482   15
5. Al-Idrisi's circular world map, 1553   16
6. Tripartite world map in twelfth-century manuscript "Etymologiae," by St. Isidore of Seville   18
7. Lazaro Luis, portolan manuscript chart of Atlantic Ocean, 1563   19
8. Map of Ceylon in manuscript copy of Ptolemy's *Geography*, ca. 1480   22
9. Map of the world in Martin Waldseemuller's 1513 edition of Ptolemy's *Geography*, Strasbourg   24
10. Map of the New World in Martin Waldseemuller's 1513 edition of Ptolemy's *Geography*, Strasbourg   25
11. Double hemispheric map of the world in Ptolemy's *Geography*, Venice, 1561   26
12. Map of the New World in Ptolemy's *Geography*, Basel, 1552   27
13. Diagram of globe gores in Albrecht Durer's *Albertus Durerus*

Nurembergensis pictor huius aetatis celeberrimus, Paris, 1535   30

14. Sphere in Thomas Blundeville's *A plaine treatise of the first principles of cosmographie,* London, 1597   35
15. Diagram of armillary sphere in Peter Apian's *Cosmographicus liber,* Landshut, 1524   38
16. Diagram in Peter Apian's *Cosmographia,* Antwerp, 1581   40
17. Volvelle in Peter Apian's *Cosmographia,* Antwerp, 1581   41
18. Frontispiece of Oronce Fine's *Orontij Fine Delphinatis,* Paris, 1542   43
19. Diagram of climatic zones in Martin Borrhaus's *Elementale cosmographicum,* Paris, 1551   44
20. Volvelle in Giovanni Paolo Gallucci's *Theatrum mundi,* Venice, 1589   45
21. Title page in Robert Record's *The Castle of Knowledge,* London, 1556   46
22. View of Anglia in Hartmann Schedel's *Liber cronicarum,* Nuremberg, 1493   50
23. World map in Sebastian Munster's edition of *Ptolemy's Geography,* Basel, 1552   53
24. Regional map in Sebastian Munster's edition of *Ptolemy's Geography,* Basel, 1552   54
25. Illustration in Andre Thevet's *Les Singularitez de la France antarctique,* Paris, 1557   57
26. Illustration in Andre Thevet's *La Cosmographie universelle d'Andre Thevet, cosmographe du Roy,* Paris, 1575   58
27. Illustration in Thomas Blundeville's *A plaine treatise of the first principles of cosmographie,* London, 1597   63
28. Illustration in *The Geomancie of Maister Christopher Cattan,* London, 1591   64

29. The grid and art; illustration in Albrecht Durer's *Albertus Durerus Nurembergensis pictor huius aetatis celeberrimus,* Paris, 1535   69

30. The grid and cartography; illustration in Albrecht Durer's *Albertus Durerus Nurembergensis pictor huius aetatis celeberrimus,* Paris, 1535   69

31. Map of the world in Abraham Ortelius's *Theatrum Orbis Terrarum,* Antwerp, 1570   73

32. Frontispiece of Abraham Ortelius's *Theatrum Orbis Terrarum,* Antwerp, 1570   75

33. Map of Iberia in Abraham Ortelius's *Theatrum Orbis Terrarum,* Antwerp, 1570   77

34. Detail from map of Asia in Abraham Ortelius's *Theatrum Orbis Terrarum,* Antwerp, 1570   78

35. Frontispiece of Mercator's *Atlas,* Duisberg, 1595   81

36. Decorative page in Mercator's *Atlas,* Duisberg, 1595   82

37. Detail from map of Africa in Mercator's *Atlas,* Duisberg, 1595   83

38. The city of Limoges from Maurice Bouguereau's *Le Theater Francois,* Tours, 1594   86

39. Map of Dorset in Christopher Saxton's *Atlas of England and Wales,* London, 1579   89

40. Detail from map of Denbeigh in Christopher Saxton's *Atlas of England and Wales,* London, 1579   90

41. Frontispiece of Michael Drayton's *Poly-Olbion,* London, 1613

42. John Norden's manuscript map of Essex, 1595   93

43. Table of distances in John Norden's *An Intended Guyde for English Travailers,* London, 1625   95

44. Map of Cambridgeshire in John Speed's *The Theatre of the Empire of Great Britaine,* London, 1611–12   98

45. Marcus Gheeraerts, *Portrait of Queen Elizabeth,* ca. 1592   100

46. Leonardo da Vinci's *Map of Imola*, ca. 1502   102
47. View of Lyon in Hartmann Schedel's *Liber cronicarum*, Nuremberg, 1493   103
48. Map of Groningen in Georg Braun's *Civitates Orbis Terrarum*, Cologne, 1612   106
49. Detail from map of Algiers in Georg Braun's *Civitates Orbis Terrarum*, Cologne, 1612   108
50. Portolan map of Southern Africa in Vallard Atlas, Dieppe, 1547   113
51. Map of Atlantic in Battista Agnese's *Portolan Atlas*, Italy, ca. 1544   114
52. Map of Eastern Mediterranean in Francesco Ghisolfi's *Portolan Atlas*, Genoa, ca. 1553   115
53. Frontispiece of *Mariners Mirrour*, London, 1588, an English translation of Janszoon Waghenaer's *Spieghel der Zeevaerdt* (1584–85)   118
54. Detail from map in Lucas Janszoon Waghenaer's *Spieghel der Zeevaerdt*, Leiden, 1584–85   119
55. Frontispiece of Willem Blaeu's *The Light of Navigation*, Amsterdam, 1612   121
56. Coastal profiles in Willem Blaeu's *The Light of Navigation*, Amsterdam, 1612   121
57. Page from *The Arte of Navigation*, London, 1584, Richard Eden's English translation of Martin Cortes's *Arte de Navegar* (1551)   123
58. Title page from *L'Arte del Navegar*, Venice, 1555, an Italian translation of Pedro Medina's *Arte de Navegar* (1545)   124
59. Title page of Edward Wright's *Certaine Errors in Navigation*, London, 1610   127
60. Gridded figure in Peter Apian's *Cosmographia*, Antwerp, 1581   130

61. Illustration of surveyor's instruments and techniques in Silvio Belli's *Libro del misurar con la vista,* Venice, 1565   131
62. Pictorial estate map of Wotton Underwood, ca. 1564–86   133
63. Page from Thomas Digges's *A Geometrical Practise, named Pantometria,* London, 1571   134
64. Map of Bagshot Park in John Norden's *A Description of the Honor of Windsor,* 1607   137
65. Title page from Aaron Rathborne's *The Surveyor,* London, 1616   139
66. Diagrams in William Leybourn's *The Compleat Surveyor,* London, 1674   140
67. Map of Virginia in Theodor de Bry's *Admiranda narratio,* Frankfurt, 1590   144
68. Detailed map of Virginia in Theodor de Bry's *Admiranda narratio,* Frankfurt, 1590   145
69. The *Relacione Geográfica* map of Tenanpulco and Matlactonatico, 1581   148
70. Surveying instruments in Nicholas Bion's *Traite de la construction et des principaux usages des instruments de mathematique,* Paris, 1709   152
71. Mapping instruments in Nicholas Bion's *Traite de la construction et des principaux usages des instruments de mathematique,* Paris, 1709   152

ACKNOWLEDGMENTS

I have been very fortunate. The writing of this book would have been impossible without access to rare charts and maps. I was privileged to receive a number of fellowships that enabled me to spend time at libraries with exceptional collections. In 1999 I was awarded a Dibner Fellowship at the Smithsonian Institution Libraries, which gave me three months at the Dibner Library of Science and Technology in Washington, D.C. I was helped enormously by Ron Brashear, Bonnie Sousa, and Bill Baxter.

In 2001 I took up a Frank Hideo Kono Fellowship to the Huntington Library in San Marino, California. Roy Ritchie, the director of research, was gracious and welcoming; people at the readers' services were accommodating; and the staff of the Ahmanson Reading Room provided helpful and unfailing assistance.

In spring 2001 I made a return visit to the library of the American Philosophical Society in Philadelphia. Roy Goodman showed his usual unflagging enthusiasm. I have also called upon the vast resources of the Library of Congress and the Folger Library, both in Washington, D.C.

The Maxwell School at Syracuse University awarded me an Appleby-Mosher Faculty Award, which allowed me to travel to the British Library in May 2001. The Faculty of Arts and Sciences at Syracuse University awarded me research leave in the spring semester of 2001, giving me the time necessary to visit collections and complete a first draft of this book.

In 2002 I took up the position of chair of the Department of Geography and Environmental Systems at the University of Maryland, Baltimore County (UMBC). Moving house, job, and metro area all at the same time could have been much more traumatic than it was, but the help and guidance of all my colleagues at UMBC gave me the support I needed to finish the book.

My really good fortune is Lisa. I only hope I have been as supportive to her in her work and life as she has been to me in mine. I dedicate the book to her, knowing full well it involved too many days in distant libraries and too much travel to faraway places.

MAKING SPACE

1

INTRODUCTION

The National Gallery in Washington, D.C., holds two paintings by the Venetian artist Sebastiano del Piombo (1485–1547). He is less well known to the general public than is Michelangelo or Raphael, but his relative obscurity is recent; he was a popular and well-respected painter in his time. His secular and religious paintings were influential, especially in Spain and Portugal, where he was more highly regarded than Michelangelo.

The two paintings that hang in Washington are portraits. The first, *Portrait of a Humanist* (ca. 1520), shows a scholar, probably the renowned poet Marcantonio Flamini, seated next to a table. On the table at his right elbow sit objects and symbols of contemporary learning, among them a set of books and a globe. The second painting is del Piombo's 1516 portrait, *Cardinal Bandinello Sauli, His Secretary, and Two Geographers*. The title says it all. The cardinal is easily identifiable: he is the only one seated and is wearing the red apparel of his office and the cardinal's cap. He returns the viewer's gaze with a look of enigmatic boredom. To his right a secretary, hunched over in obeisance, whispers in his ear; to the cardinal's left, two men, geographers according to the title, engage in some form of argument, a disputation perhaps about the text in front of them. The cardinal's left hand draws our eye to a rug-covered table, on which sit a bell and an open book, the latter containing maps and writing. The book is a copy of Ptolemy's *Geography*.

The picture has a host of interesting details. On the cardinal's knee sits a small fly. In earlier times the fly was a symbol of corrup-

Making Space

1. Sebastiano del Piombo, *Cardinal Bandinello Sauli, His Secretary, and Two Geographers*, 1516, oil on canvas. By permission of the National Gallery of Art, Washington, D.C.

tion but by the time of the Renaissance it was often used in paintings as a decoy to ward flies off while the paint dried. However, the symbolic meaning comes to the fore for the modern viewer, who knows that, while this portrait was taken at the height of Cardinal Sauli's power, five years after he was made a cardinal by Pope Julius, the next papacy found his fortunes waning. He was imprisoned in 1517 for plotting against Pope Leo X. We are looking at the portrait of a man whose bright star of ambition, so finely revealed in the portrait, was about to be extinguished.

I mention these two paintings because, in their different ways, they highlight a central feature of the Renaissance world. The humanist's globe and the cardinal's atlas were not just standard props used by a clever painter; they were symbols of a radical representation of the world that embodied a new way of seeing. It is this way of seeing that I want to explore.

Making Space

The space in which most of the contemporary world is viewed, a gridded space empty of history yet full of promise, was constructed in Europe between 1475 and 1600. This period marks a transition zone between two differing views of space. The first view saw space and time as being deeply intertwined, with history as well as geography forming an important part of geographical representation. The second view, however, began to visualize a space more independent of history. These views do not constitute a simple dichotomy between an easily demarcated "before" and "after." The medieval period was not an unchanging block, and the Renaissance was deeply marked by a medieval heritage. However, it is possible to discern the construction of a modern space of the grid, the map, and the survey. Plotted on the grid of latitude and longitude, this new world was produced on maps and negotiated in new methods in navigation and surveying that allowed the world to be not only seen but also explored and appropriated.

Mapping was never politically neutral or socially indeterminate. The Renaissance introduced a new way of seeing the world, describing the world, and mapping the world that anticipated both the Enlightenment and colonialism. Maps of the Renaissance reflect and embody new forms of scientific understanding and new techniques of territorial appropriation.

Six Spatial Discourses

In this book I will look at six spatial discourses: the construction of the grid; the emergence of cosmography; the mappings of the world; the navigation of the oceans; the surveying of the land; and the annexing of colonial territories.

We see the world through a grid that was first developed in the classical world. The Renaissance coordinating of the world, its depiction on a grid system, stems directly from the translations of Ptolemy's *Geography*. Ptolemy was a Greek-Egyptian who worked in Alexandria in the second century C.E. His work, especially his geographical work and maps, came into Western Europe through the efforts of Arabic scholars who translated and modified it. In *Geography,* Ptolemy discussed suitable map projections for a world map and provided a table of latitude and longitude for the cities of the known world. The first Latin translation of his work appeared in manuscript form in the early fifteenth century, but after 1475 almost thirty different printed versions were published in a variety of languages. In successive editions, editors would publish both Ptolemy's original findings and contemporary maps. The Ptolemaic legacy was the grid of latitude and longitude that plotted the places of the world.

I will also look at the encompassing view of the world that developed in sixteenth-century cosmography. A synthesis of astronomy, geography, and astrology, cosmography covered a broad range of subjects, creating, in essence, a theory of how the terrestrial and celestial realms were connected. Of this range, I will focus on three areas: the technical gaze, cosmo-geographies, and occult discourses. I will examine the cosmographical work of Peter Apian (1495–1552) and Oronce Fine (1494–1555). Both men wrote works on cosmography that spanned geography, astrology, surveying, and cartography. By looking at their work in detail, we can see the medieval basis of much Renaissance scholarship, as well as the beginnings of new forms of scientific instrumentation and measurement of the world. Both scholars developed new cartographic projections, made world maps, and devised new instruments to "see" the world. Their view of a coherent universe provided a model of the world for more than one hundred years. Apian's most famous work is *Cosmographia,* which was first published in 1524 and went through numerous translations and editions throughout the sixteenth century. Fine served as cosmographer to the kings of France and published a variety of works. Both Apian and Fine carried on a vigorous correspondence with scholars around Europe, so an understanding of their work allows an understanding of the intellectual construction of the world in the European Renaissance.

An important part of cosmography developed in the sixteenth century: a body of geographical writings that referenced the medieval world but also prefigured the modern world. The cosmogeographers I consider are Sebastian Munster (1488–1552), Andre Thevet (1516–92), and John Dee (1527–1608). These three men sought to construct a geographical understanding of a world. But it was a world growing in size and complexity as Europeans expanded their connections with a wider world. I study the first two writers to explore the paradox of a project that sought to impose order on an unruly world, while I examine John Dee to explore the occult dimensions of Renaissance cosmography.

I will also explore the mappings of the world that emerged in the sixteenth century in world atlases, national atlases, and city atlases. In a truly remarkable fifteen-year period, the first modern printed atlas appeared (Ortelius's *Theatrum* in 1570), then the first printed urban atlas (Braun and Hogenberg's *Civitates* in 1572), and finally the first printed national atlas (Saxton's *Atlas* in 1579). These works condense the Renaissance mapping of the world. I focus on the work of two cartographers: Abraham Ortelius (1527–98), whose *Theatrum Orbis Terrarum* is arguably one of the world's first atlases and without question one of the most beautiful and influential; and Gerardus Mercator (1512–94), whose 1595 atlas is a massive work of geographic compilation and cartographic representation. Urban mappings reached their peak in the city maps of Braun and Hogenberg's *Civitates orbis terrarum* of 1572–1618. In a discussion of "national geographies," I will also look at Christopher Saxton's 1579 atlas of England and the subsequent work of the Elizabethan chorographers and mapmakers John Speed and John Norden.

The grid that Ptolemy outlined had to be more finely calibrated in order to survey the world and chart the oceans. William Cuningham's 1559 book *The Cosmographical Glasse* was subtitled "the pleasant principles of cosmographie, geographie, hydrographie or navigation." In the latter half of the sixteenth century, these subtitled fields would emerge as distinct fields of technical inquiry. I will look at the navigation of the oceans and the surveying of land, both of which were rudimentary in the middle of the century. Pedro Medina's 1555 *Arte del Navegar* (Art of navigation) and Belli's 1565 *Libro del Misurar* (Book of surveying) show simple tools and limited techniques. But by the end of the century, more sophisticated ways of measuring and navigating the world had been developed. Lucas Waghenaer's *Mariner's Mirrour* of 1584–85 and Aaron Rathborne's 1616 *The Surveyor* exemplify these more sophisticated spatial discourses.

The techniques of navigation, mapping, and surveying were always tied to a larger purpose. In the sixteenth century we can clearly see the emergence of colonial cartographies as navigation was used to find new lands and the techniques of spatial surveillance were used to map and appropriate new territories. In chapter 7, I provide two examples of such uses, one drawn from the very early English experience in North America and the other from the Spanish empire in Central America.

This work is part of a larger cartographic project. In my book *Representing the Republic* (2001), I deconstructed the cartographic representation of the United States from 1600 to 1900 and presented a longer theoretical discussion of the meaning of the map. In *Alternative Geographies* (2000), I placed cartographic representations in a wider arc of knowledge production so that Mercator is seen less as a forerunner of the scientific revolution and more as a Renaissance magus.

The dates in the subtitle need some explanation. The year 1475 marks the first printing of Ptolemy's *Geography*. The end date of 1600 is an approximation. Some of my analysis will spill over; the date is more of a ragged break than a sharp edge.

The basic argument of this book is that modern space—the space the modern world inhabits and "sees"—was created in Europe between 1475 and 1600. It was produced using a variety of means, including the use of the grid to plot the world; the use of the cosmographical sphere as the starting point for the mathematically derived practices of navigation and surveying; the increasing use of maps; and the creation of a cartographic language for new mappings of the world, states, and cities. In this new spatial practice, the world was enmeshed in a grid, laced with compass lines and seen through the lens of the theodolite, back-staff and cross-staff. New techniques of spatial surveillance were employed by the state, private companies, and powerful individuals in acts of land commodification and colonial appropriation.

But we would make a mistake if we saw the new spatial practice as a sort of proto-science that emerged untouched by the medieval worldview. Alchemy and astrology were as important as mathematics and astronomy. The early cosmographers were astrologers and alchemists as well as mathematicians and instrument makers; they used numbers as mystical intimations as much as instrumental values. The distinctions that we make today between what is and what is not science were not so sharp. An interest in astrology did not preclude a concern with mathematical accuracy in navigation.

The new spatial sensitivity was apparent in the literature of the day. By the late sixteenth and early seventeenth centuries, writers could assume a spatial awareness among their readers, as is seen in the rich imagery of John Donne's poems written between 1593 and 1601:

> On a round ball
> A workeman that hath copies by, can lay
> An Europe, Africa, and an Asia
>   "A Valediction: of Weeping"

Another Donne poem from the same time, "A Valediction: Forbidding Mourning," uses complex imagery of a compass to refer to two lovers: when one returns home, the other one stands tall. The poem is all the more satisfying as a metaphorical musing on compasses and navigation when you know that—in a model of the 360 degrees of complete circumference—the poem has precisely thirty-six lines and ends with the line, "And makes me end, where I begun."

In one section of Edmund Spenser's 1590 epic poem, *Faerie Queen,* the techniques of navigation become a source of poetry:

> Upon his card and compass firms his eye,
> The maysters of his long experiment,
> And to them, does the steddy helme apply,
> Bidding his winged vessell fairley forward fly

In addition to its use in rhetorical and literary devices, spatial revisioning was deeply connected to the promotion of commercial and national interests. Developments in navigation, for example, were often encouraged and promoted by both the state and merchant companies. Surveying was intimately connected to the rising cost of land and the increasing commodification of agriculture. The steady rise in land prices in the sixteenth century was an important factor in the increasing use of the spatial practices of mapping and surveying.

The spatial discourses of the sixteenth century formed a remarkable revolution that changed the way we represent the world. The cosmos was bound in a sphere, the world was gridded and plotted, the seas were navigated, and the land was surveyed. Spatial practices were codified, a spatial sensitivity was created, and a cartographic literacy was established. The medieval flat text was replaced with the modern conception of looking on the world through gridded windows and optical instruments. We see the world through eyes that came into focus during the sixteenth century.

The European Renaissance was a revisioning, a reorientation in how we view and represent the world. It was a transformative gaze whose legacy lives on today. At the beginning of the third millennium, we still look out on the world with a gaze that was first developed in the European Renaissance. Modern space and the space of the modern were made in the sixteenth century.

2

COORDINATING THE WORLD

Hereford Cathedral in west central England is home to one of the largest maps made in medieval Europe. It was made some time around 1300 and is attributed to one Richard of Holdingham, a priest from Lincolnshire. The map measures 1.58 meters by 1.33 meters. It is a parchment map of the world. At first glance, it is not easily distinguished as a map of the world. It is circular, with an odd combination of land and water, and in the margins are strange signs and symbols. Jerusalem is at the center of this world. To our modern eye it does not look much like a map of the world.[1] Compare it with the map, shown in illustration 9, which is almost five hundred years old but looks more like a world map. These two maps were produced on opposite sides of the spatial revolution that is the subject of this book.

The revisioning that made up this spatial revolution involved a concern with the possibilities of space rather than with the constraints of time, the decline of the religious imagination, and the rise of an ostensibly scientific discourse. There are many points of difference between the two maps, but, for the moment, let us look at one that you probably have not considered: one is displayed on a grid and one is not. In essence, the Renaissance involved a gridding of the world.

The grid is a system of vertical and horizontal lines that structures the visioning and representation of space. Locating objects on the latticelike pattern is now such an accepted part of our spatial organization of the world that we take it for granted. Yet

2. The Hereford Cathedral Mappa Mundi, crayon, ink, and paint on vellum, ca. 1300. © The Dean and Chapter of Hereford and the Hereford Mappa Mundi Trust.

it has not always been so. The grid had to be invented. To see the world enmeshed in a grid implies a sophisticated mathematics. But more than an understanding of mathematics is involved; the grid implies a conceptual ability to imagine the world in purely spatial terms. By simplifying the representation of the world to the abstractly geometric, a major revolution was wrought. The grid reimagines the rough terrain of the world into a pure geometry; place is transformed into space; physical thereness is replaced by a graphical location. The grid allows the world to be plotted and replotted, mapped and remapped, represented and reimagined.

Constructing the Grid

One of the most important figures in this spatial revolution was Claudius Ptolemy. He was a Greek-Egyptian who worked in the great Library of Alexandria in the second century after the birth of Christ. We know very little about him. He lived sometime between 127 and 145 C.E. Where he was born, what he looked like, and his exact connection to the work that bears his name are not fully known. We know a bit more of where he was. He spent his adult life in Alexandria. It was a Greek city, founded in 331 B.C.E. and named after Alexander the Great. Alexander told his architect Dinocrates to build the city on a strip of land between the Mediterranean and Lake Mareotis. After Alexander's death, the city became the capital of an empire ruled by one of Alexander's generals, who called himself Ptolemy and created a dynasty over what is now Egypt that lasted seven generations, until the reign of Cleopatra. The city became one of the wonders of the classical world, competing with Rome and Constantinople for size and grandeur. Its population grew to almost one-half million; it was a major trading center, its giant lighthouse forming a beacon to the many ships that plied their trade around the Mediterranean. There were palaces and large public buildings, including assembly halls, gymnasiums, and bath houses. At the heart of the intellectual life of the city was the Great Library. The early Ptolemys, especially Ptolemy the First (366–283 B.C.E.) and Ptolemy the Second (308–246 B.C.E.), established a library that eventually held more than seven hundred thousand volumes. The Ptolemys collected a great number of manuscripts for the library: every ship that passed through Alexandria had to surrender any on board manuscripts for copying. The library obtained the major Greek tragedies from Athens, purchased the library of Aristotle, and collected Buddhist texts from India.

The library was a major center of scholarship and the intellectual hub of the Hellenistic world. Manuscripts were copied, and the Alexandrian editions became the standard editions. Dictionaries and rules of grammar were constructed. Much of Greek and indeed Babylonian learning was transcribed, discussed, amended, systematized, and improved. Jewish scholars translated the Torah from Hebrew into Greek. It was at the library that the astronomer Aristarchus argued that the earth revolved around the sun and Euclid completed his *Elements*. One of the chief librarians, Eratosthenes, assembled information sent back by explorers and travelers into a world map that was to lay the basis for Ptolemy's work. For a scholar and a writer, there was no better place in the world to be. Ptolemy was able to draw upon a vast pool of classical learning.

A word of caution about when we use the term *Ptolemy* with reference to specific works. He is not a writer whose works have come down to us untouched from the original starting point of Alexandria. His work was translated into Arabic from Greek, and from Arabic and Greek into Latin, and then into many other languages. These were not just simple acts of translation; they involved creative editing and compilation, additions and subtractions. When we use the term *Ptolemy* we are really referring to a transmission belt along which many hands have added things to the "original" message. Ptolemy's name covers the work of a variety of scholars—some known, many unknown.

The basic mathematics for Ptolemy's work had been figured by Euclid in his *Elements* around 300 B.C.E. In his famous text, Euclid gave definitions of points, lines, triangles, circles, and other basic shapes. He then outlined ten basic axioms, such as the fact that it is possible to draw a straight line between any two points. From these deceptively simple building blocks, Euclid constructed theorems, one of which I still remember from my school days: the base angles of an isosceles triangle are equal.

Euclid created an abstract space, a pure world of lines and points, triangles, circles, and spheres. This rational geometry was then mapped onto the physical world. The root definition of *geometry* is "measuring the earth" (*geo* = "earth"; *metry* = "measure"); the term shares the same "earth" definition as *geography* and *geology*. Euclid replaced the coarse place of the physical world with the space of perfect geometry.

Ptolemy's *Geography*

Ptolemy used this geometry in his work on astronomy and geography. The *Almagest* is his earliest major work. The title comes from the Arabic for "great compilation." In this thirteen-volume encyclopedia, Ptolemy brings together existing Greek knowledge into a systematized whole with distinctive coherence. Book 1 outlines a planetary model of a stationary spherical earth around which the fixed stars revolve from east to west. He develops a trigonometry that allows him to examine the annual variation in solar declination. Book 2 provides a table of rising time at various latitudes. This knowledge is an essential astrological requirement for creating horoscopes. Book 3 provides estimates of the length of the year. Books 4 and 5 deal with the moon and its movement across the night sky. Books 7 and 8 provide tables of the latitude and longitude of more than one thousand stars. Books 9 through 13 discuss the planets. In an elegant trigonometric formulation, Ptolemy plots the location of the sun, moon, stars, and planets along with their trajectories across the sky. The *Almagest* is a map of the heavens.

If the *Almagest* maps the celestial universe, then Ptolemy's *Geography* maps the terrestrial world. The work, written between 127 and 155 C.E., is in eight books. Book 1 is a general description of latitude and longitude and their measurement. Ptolemy draws on the work of earlier writers; Eratosthenes (ca. 275–194 B.C.E.) and Hipparchus (ca. 190–126 B.C.E.) had already developed the idea of imaginary lines drawn across the surface of the earth. These lines were based on the older Babylonian number system of sixty. Lines of latitude parallel to the equator were divided into degrees and minutes, with 0 degrees at the equator and 90 degrees north at the North Pole. Lines of longitude were divided into 180 degrees east and west of a prime meridian, which Ptolemy sets in the Fortunate Islands, near the present-day Canary Islands.

Ptolemy casts a grid over the surface of the habitable world that allows places to be identified by their coordinates of latitude and longitude. As he wrote in Book 1, "We are able therefore to know at once the exact position of any particular place; and the position of the various countries, how they are situated in regard to one another, how situated as regards the whole inhabited world."[2]

The grid is a subtle device that allows the world to be twisted and turned, expanded and shrunk. The lines of latitude and longitude could be bent into all kinds of wonderful shapes. We call these shapes map projections, and in Book 1 Ptolemy outlines two projections.

Books 2 through 7 contain tables of the latitude and longitude of sites in different parts of the known world, including Europe, Africa, and Asia. In Book 8 Ptolemy makes some general com-

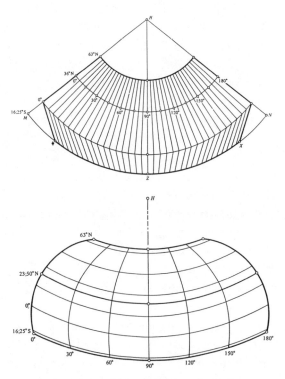

3. Ptolemy's world map projections; after J. P. Snyder's *Flattening The Earth: Two Thousand Years of Map Projections* (Chicago: Univ. of Chicago Press, 1993).

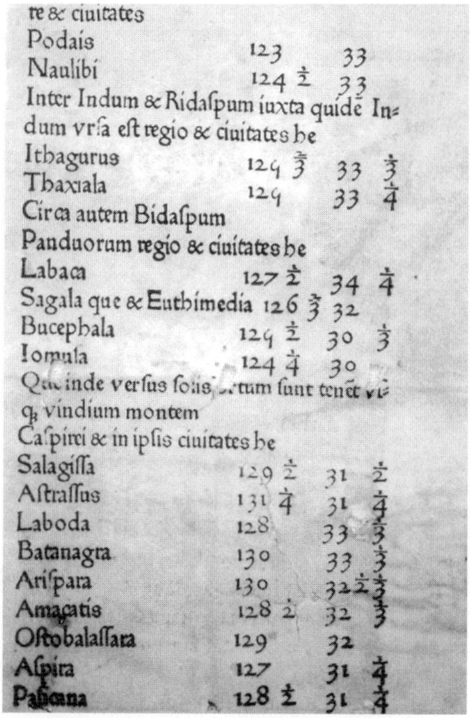
re & ciuitates
Podais 123 33
Naulibi 124 ½ 33
Inter Indum & Ridaspum iuxta quidē Indum vrsa est regio & ciuitates he
Ithagurus 124 ⅓ 33 ⅓
Thaxiala 124 33 ¼
Circa autem Bidaspum
Panduorum regio & ciuitates he
Labaca 127 ½ 34 ¼
Sagala que & Euthimedia 126 ⅓ 32
Bucephala 124 ½ 30 ⅓
Iomula 124 ¼ 30
Q… inde versus solis …rtum sunt tenet vi…
q; vindium montem
Caspirei & in ipsis ciuitates he
Salagissa 129 ½ 31 ½
Astrassus 131 ¼ 31 ¼
Laboda 128 33 ⅓
Batanagra 130 33 ⅓
Arispara 130 32 ½ ⅓
Amacatis 128 ½ 32 ⅓
Ostobalassara 129 32
Aspira 127 31 ¼
Pasicana 128 ½ 31 ¼

4. Latitude and longitude in Ptolemy's *Geography*; detail from a 1482 manuscript fragment in the Library of Congress, Washington, D.C. Courtesy of the Library of Congress.

ments on map construction and lists ten maps for Europe, four for Africa, and twelve for Asia. These maps did not survive, but Ptolemy had provided the code of how to make them. Ptolemy provides a general model as well as specific coordinated data that allowed later scholars to make world and regional maps from his data.

Ptolemy's influence quickly disappeared from the Western imagination. His work, however, was kept alive in the Arab world. In the seventh and eighth centuries C.E., Arabs had conquered most of the ancient world, from the Middle East west along North

Africa and up to Spain, east toward India, and north up to Armenia. This political empire was also a scene of intellectual activity.

Ptolemy's work was translated by the Arab cosmologists of the Abbasid court in Baghdad around 800 C.E. Al-Battani (ca. 880) restated the *Almagest* with improved measurements, while *Geography* was revised by al-Khwarizma around 820 in his *Face of the Earth*, which included a map of the world.

The translation and improvement of Ptolemy's work was part of a broader intellectual renaissance. In many cities, scholars—Christians and Jews as well as Muslims—translated the work of the ancients and worked on new themes. Al-Idrisi (ca. 1100–1166 C.E.), for example, was born in Ceuta, Morocco. He studied at Cordoba in Spain and was employed by a Norman King, Roger the Second of Sicily. Al-Idrisi was a practicing geographer who traveled through the Arab world and retraced the paths of the earlier Islamic conquest. He constructed a celestial sphere and a map of the world. Al-Idrisi was part of a larger geographical school of

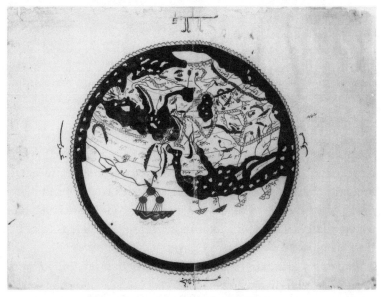

5. Al-Idrisi's circular world map, 1553; from MS Pococke 375, ff. 3v–4r. By permission of the Bodleian Library, University of Oxford.

scholarship in the Arab world that involved geographical fieldwork and travel as well as book knowledge, including familiarity with Al-Yaqubi's *Book of Countries* (ca. 891), Al-Balkhis's *Figures of the Climates* (921), Al-Muqaddasi's *The Best Description for an Understanding of All Provinces* (ca. 985), and Al-Bakri's *Book of Roads and Kingdoms* (ca. 1050).

European Maps Before the Grid

While Ptolemy's work was being developed in the Arab world and before it diffused throughout Europe, medieval Europe represented the world in two important ways. First, there were medieval maps of the world, *mappaemundi*. The term *mappaemundi* derives from the Latin *mappa,* meaning "cloth or napkin," and *mundus,* "the world." These maps were often very small and integral parts of manuscript texts, often appearing in the margins, surrounded by writing. Three main types have been identified: tripartite, zonal, and transitional.[3] The most common were the tripartite maps, which typically centered on Jerusalem, placing Asia at the top, Africa in the east, and Europe in the west. We still have as legacy the term *orientate,* which comes from the term *orient* for "east." Tripartite maps are sometimes referred to as T-O maps, because their outline combines a *T* shape inside an *O* shape. *Mappaemundi* were not merely representations of the physical world; they were also symbolic objects. The tripartite maps, for example, symbolized the division of the world by God to the three sons of Noah: Shem, Ham, and Japheth. The T-O maps have also been seen as representing the Christian cross. Thus the *mappaemundi* were cosmologies that reflected a deeply religious view of the world. While many of these *mappaemundi* were small and could be found incorporated into texts, a few were freestanding, such as the Hereford world map; they were large artifacts meant for public display. The cartographic equivalent of the rose window in churches, they provided a religious image of the world.

Look again at the Hereford map. As a work of geography it is perhaps easier to see if we imagine it as centered on the Mediter-

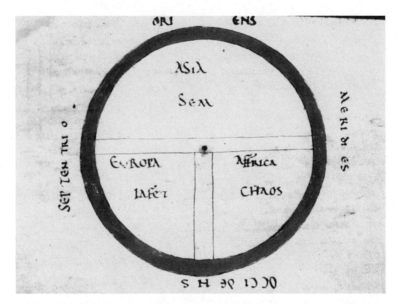

6. Tripartite world map in twelfth-century manuscript "Etymologiae," by St. Isidore of Seville. By permission of the Bibliothèque nationale de France, Paris.

ranean with east at the top and the British Isles at the bottom left. The cities of Europe are accurately located. But it was also a work of a religious imagination: Christ at the Last Judgement appears at the top of the map and the letters "MORS" are located around the border to remind us of the finitude of our life on earth. It is a work of religious history as well as geography: scenes on the map include the crossing of the Red Sea, the tower of Babel, and the expulsion of Adam and Eve. Medieval world maps represented time as well as space, history as well as geography, and religious doctrine as well as topography. Medieval *mappaemundi* were works of religious iconography as well as geographic representations of the world.

A second medieval form of representation was the portolan map. These maps began to appear in Europe in the thirteenth century. They got their name from *portolano*, the Italian word for the

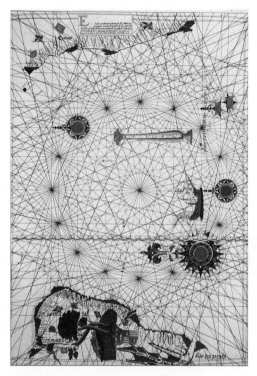

7. Portolan manuscript chart of Atlantic Ocean by Lazaro Luis, 1563. Staatsbibliothek, Dresden.

pilot book that contained written sailing directions and information about courses, anchorages, and ports. The portolans were initially maps for sea travel. They developed out of the invention and refinement of the compass. Portolans were covered with a network of lines corresponding to the principal points of the compass. Seafarers used the lines on the chart to set a course. Portolan maps were the seafarers' view of the world; the emphasis was on coasts and ports rather than on interiors. Although many portolans had latitude in their borders, the grid was not formalized in the early portolans. In both the *mappaemundi* and the earlier portolan maps there was no standard grid. The world was held together by religious history or local lines of wind and compass directions. The full gridding of the world had to await the rediscovery of Ptolemy.

Ptolemy and the Renaissance

Ptolemy was a major figure of the European Renaissance; he was published and printed, copied and amended, written and read by artists, humanists, scholars and explorers, princes and priests, merchants and prelates. He was depicted in Raphael's 1510 *School of Athens* as one of the great intellects, and he frequently appeared in the borders of world maps and even in sculptures. On the Campanile of the Duomo in Florence, there is a relief of him, by Giotto and Andrea Pisano, sitting at a desk, looking at a globe, and holding an astrolabe in hand.

Ptolemy's *Geography* was hugely influential: its translation was a central part of the European Renaissance. It was also a material object of great monetary and symbolic value, whose purchase and ownership was a form of conspicuous consumption. Its appearance in Piombo's portrait of Cardinal Sauli speaks of luxury as well as learning.

There are manuscript copies of Ptolemy's *Geography* dating from as early as the thirteenth century. These Byzantine copies, written in Greek, came in a variety of forms; some contained maps, some did not. Those with maps came in two main types: the A version contained twenty-seven maps while the B version had sixty-five maps. It is unlikely that Ptolemy actually drew these maps; more likely, they were the work of later scribes.

Ptolemy's writings, especially his *Almagest* and *Tetrabiblos*, were known in Western Europe starting around 1200. But by the early 1400s, Byzantine scholars and their Greek manuscripts, including manuscripts of Ptolemy's writings, were brought to Italy. Greek was little known in Western Europe, where even most scholars understood Latin rather than Greek; so the texts had to be translated. A translation of Ptolemy's *Geography* was begun by Emanuel

Chrysoloras and the text was finally translated into Latin by Jacopo d'Angiolo of Tuscany in 1406. This text was widely distributed.

Before printing, of course, manuscripts were copied by hand. In the case of Ptolemy's *Geography*, as with other texts, the various translations and different copies involved changes and amendments rather than straightforward translations. In 1427 the Canon of Rheims in France, Guillaume Fillastre, told the copying scribe to add a map of Northern Europe by the Danish geographer Claudius Clavus. In 1466, Donnus Nicolaus Germanus presented the Duke of Ferrara with a copy of *Geography* that included new map projections and a unique cartographic key for representing physical features and boundaries. The Florentine painter Pietro del Massaio designed two manuscript copies of *Geography* in 1469 and 1472 that contained contemporary maps of Spain, France, and Italy as well as perspective views of the major cities of the Mediterranean. Producing a manuscript copy of Ptolemy's work was less a mindless act of copying and more a creative opportunity to generate new maps and projections.

Ptolemy's *Geography* appeared as a number of manuscript atlases in the late fifteenth and early sixteenth centuries. One extant example is the Wilton Codex, now in the Huntington Library in San Marino, California.[4] This manuscript atlas contains twenty-seven maps. The world map in the Wilton Codex was printed, rather than hand drawn, on vellum, based in all probability on a printed version produced in about 1486. The printed map in this manuscript atlas indicates the close connection between printed and hand-drawn texts at the end of the fifteenth century.

A history of the printing of Ptolemy's *Geography* encapsulates the early history of printing in Europe. The first printed copy appeared in Vicenza in 1475. It contained no maps but had two diagrams of map projections. At about the same time other editions appeared in Rome (1477), Florence (1480), and Ulm (1482). The Ulm edition, with its painted roundels and richly colored woodblock maps with deep blue seas and yellow borders, is arguably one of the most beautiful.

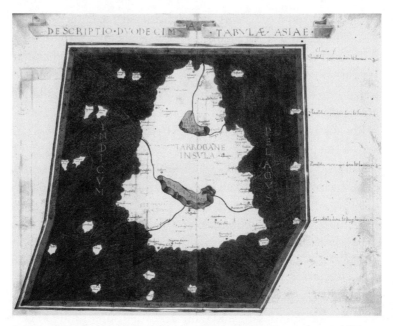

8. Map of Ceylon in manuscript copy of Ptolemy's *Geography*, ca. 1480; HM 1092, Wilton Codex, ff. 53v–54r. This item is reproduced by permission of The Huntington Library, San Marino, California.

More than fifty editions appeared between 1475 and 1730 in a variety of forms (see appendix). The editions were published in many European cities including Amsterdam, Arnhem, Basle, Bologna, Cologne, Cracow, Douay, Frankfurt, Leyden, London, Louvain, Lyons, Padua, Strasbourg, Utrecht, Venice, and Vienna. The text was published in lavish editions for wealthy patrons, such as the beautiful Ulm edition, as well as in smaller, more modest books meant for wider circulation to the less wealthy, such as the 1548 and 1561 editions. The maps were produced in copperplate as well as woodcut. Most of the editions contained contemporary maps as well as Ptolemy's original maps. After 1508, editions of Ptolemy's *Geography* included maps that incorporated the discoveries of the New World. In fact, the first printed map of the New World appears in an edition of *Geography*.

Those who worked on the various editions constitute a who's who of Renaissance cartographers: Nicolaus and Jodocus Hondius, Gerardus Mercator, Sebastian Munster, Johan Ruysch, Bernardus Sylvanus, and Martin Waldseemuller. Working on Ptolemy's *Geography* was a vital part of sixteenth-century cartographic development. The Venetian mapmaker Giovanni Matteo Contarini described himself on his 1506 world map as "famed in the Ptolemaen art."

Ptolemy provided an opportunity for each generation of mapmakers to try out their skills, to test themselves against a common standard. Working on Ptolemy's text offered a new chance to represent the world. Ptolemy's *Geography* was a publishing genre that allowed scholars to work on an old theme while also developing new twists: new projections, new maps, new ways of looking at a new world. The publishing of Ptolemy was, after the Bible, one of the most important publishing ventures of the Renaissance and the primary impulse to the creative cartography of the sixteenth century. Successive editions of *Geography* in the sixteenth century also give us a picture of an expanding world. A 1513 edition by Martin Waldseemuller had twenty maps, while a 1548 edition by Giacomo Gastaldi had thirty-three. Newer editions allowed the recently discovered New Worlds to be incorporated into the world picture.

Let us look at some editions in more detail. The first edition to include maps was published by Arnold Bucknick in Rome in 1477. It was titled Ptolemy's *Cosmographia*. The project was developed by Conrad Sweynheym of Mainz, who along with Arnold Pannartz established the first printing press in Italy at Subiaco, thirty miles east of Rome, in 1464. Both were clerics in the Benedictine monastery at Subiaco, a predominantly German community. They undertook an ambitious program of printing German and Latin authors. Sweynheym persuaded the Veronese humanist Domizio Calderini to establish an improved Latin text from the various manuscript versions of Ptolemy. Sweynheym died in 1477, but the book was brought out by Arnold Bucknick, who was probably an assistant in Sweynheym's printing office. The map of the world

24 ❧ Making Space

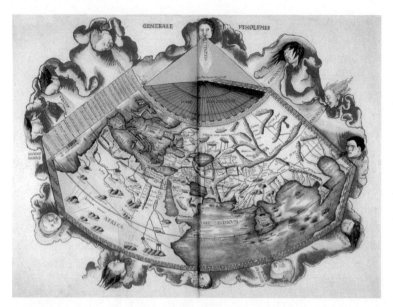

9. Map of the world in Martin Waldseemuller's 1513 edition of Ptolemy's *Geography*, Strasbourg. Courtesy of the Library of Congress, Washington, D.C.

was based on Ptolemy's first conical projection. The other maps are all trapezoid, except Taproba (Sri Lanka), which is on a cylindrical projection. All the maps show a clean precision with good detail. The plates were used in numerous editions up to 1508 with many additions including introductory chapters on mathematical geography. The 1508 edition had a modern map of the world by Johan Ruysch with a written commentary. About five hundred copies of this Rome edition were produced—a fairly successful print run for an expensive book. Christopher Columbus owned and annotated a copy of the 1478 edition, which he referred to in his letters. It is fair to say that this Rome edition was part of the worldview of both ambitious explorers and study-bound scholars. The world represented in the Rome edition was a view held by Columbus as he argued his case for support in the courts of Europe and eventually set out from Spain to sail across the ocean.

An early sixteenth-century edition was begun by Martin Waldseemuller. It was published in 1513 in German in Strasbourg. It is the Jacopo d'Angiolo's translation revised by Matthew Ringmann and edited by two lawyers, Jacob Aezler and Georg Uebelin. It is a lavish, large book, featuring expensive paper and (in the Huntington Library's copy) woodblock maps beautifully colored in rich, vibrant colors. The maps of Ptolemy were supplemented by new maps, termed *Tabula Nova* or sometimes *Tabula Moderna*, including a map of the New World showing the Caribbean islands and a small sliver of South America.

There were also smaller volumes produced for a larger audience. A 1548 edition was printed in Italian in Venice from a translation by Mattioli, who was a "learned physician of Siena." It is a small handbook with only sixty pages of text. There were twenty-six Ptolemy maps and thirty-four new maps, including the New World and more detailed maps of Italy that contained the Piedmont and Venezia. All the maps were gridded, with latitude and longitude scaled into the map borders.

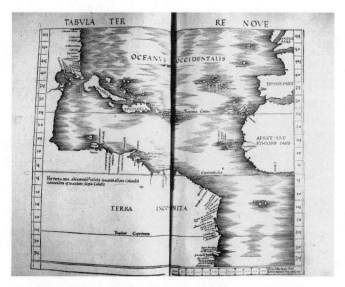

10. Map of the New World in Martin Waldseemuller's 1513 edition of Ptolemy's *Geography*, Strasbourg. Courtesy of the Library of Congress, Washington, D.C.

The vigorous Venetian book industry produced another, larger edition in 1561 that contained sixty-four copperplate maps. Ptolemy's world and regional maps were now complemented by a double hemispheric world map, which allowed a wider angle on the New World as well as *Tabula Nova* of regions of the world including Brazil and Cuba. Ptolemy had supplied a method that allowed an expanding world to be represented.

Through the course of the sixteenth century, Ptolemy's *Geography* moved from a work of major stimulus to a work of historical interest. By the 1548 Gastaldi edition Ptolemy was relegated to an appendix and, by the 1578 Mercator edition, he was treated as a figure of historical interest.

Ptolemy's *Geography* stimulated new knowledge and innovative cartographic techniques; it was a geographical treatise that provided the basis for the age of discovery and the revisioning of the world. His continuing legacy includes orienting maps with north at the top, applying the system of latitude and longitude,

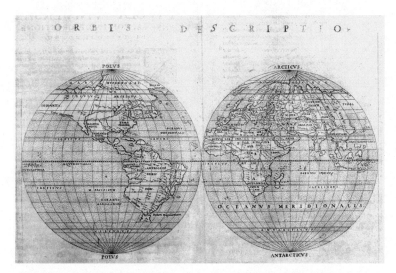

11. Double hemispheric map of the world in Ptolemy's *Geography*, Venice, 1561; RB 106421. This item is reproduced by permission of The Huntington Library, San Marino, California.

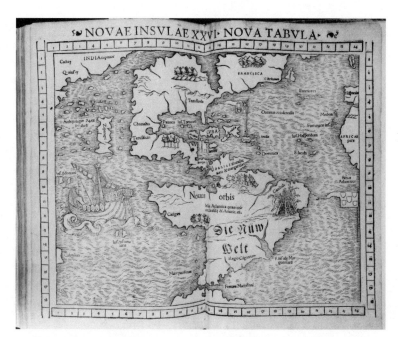

12. Map of the New World in Ptolemy's *Geography*, Basel, 1552. Courtesy of Dibner Library for the History of Science and Technology, Smithsonian Institution Libraries, Washington, D.C.

using gazetteers in maps and atlases, and using globes and a variety of map projections. Ptolemy introduced us to the idea and practice of a coordinated world.

"The habitable earth as spherical"

In *Geography*, Ptolemy noted that there were two ways to represent the earth: "one is to represent the habitable earth as spherical; the other is to represent it as a plane surface."[5] The publication and dissemination of Ptolemy's *Geography* stimulated both kinds of representations.

Although the sphericity of the earth and the heavens was known as early as 300 B.C.E., there are few examples of three-dimensional globes in Europe prior to the Renaissance. One of these examples is a classical sculpture of Atlas with celestial globe on his shoulder uses a similar vernal equinox to Ptolemy's *Almagest*. The globe was based on Ptolemy's work, so it must have been done after 150 C.E. There is also an Islamic celestial globe made in Valencia around 1080 C.E. However, terrestrial globes appear in more abundance in the wake of the translation and dissemination of Ptolemy's *Geography*.

The oldest extant terrestrial globe, made in 1492, is by Martin Behaim (1459–1507). He was a merchant in Nuremberg who had lived in Portugal and wanted to represent the commercial trade opportunities with the East and finance an expedition to China. His globe is a mixture of ancient Ptolemaic maps, medieval European maps, and portolan charts.

In 1515 a professor in Nuremberg, Johannes Schoner (1471–1547), made his first globe. Other globes followed in 1523 and 1533. All Schoner's terrestrial globes show a passage between the Atlantic and the Pacific at the tip of South America. Schoner was also the first to sell terrestrial and celestial globes as matching pairs. We have a visual record of Schoner's work; the celestial globe depicted in the famous painting *The Ambassadors*, completed around 1533 by Hans Holbein, has been verified as being one of Schoner's.

Globes were important methods of displaying and recording geographical information that was vitally important to merchants and traders venturing beyond the confines of Europe. Globe production remained a highly specialized and highly localized activity that was closely tied to centers of mercantile activity. But the centers of globe production shifted over time, in delayed response to the rise and fall of mercantile dominance: first Nuremberg, then Flanders and Amsterdam, and then Italy. Nuremberg was a center for globe production until the nineteenth century. Other notable globe makers were also active in Flanders. The inventor of triangulation and the pupil of Peter Apian, Gemma Frisius made his first globe in 1529–30 in Louvain; in 1536 he jointly produced a globe with the great mapmaker Mercator. This globe depicted the latest discoveries of the Portuguese. Mercator also made globes on his own; he made matching globes in 1541 and 1551. His globe of 1541 was to provide the basis for the map projection that takes his name.

Globe and map production shifted northwards. Petrus Plancius (1552–1622), an important globe maker from Antwerp who settled in Amsterdam, was both a cartographer and a theologian. He was also the cartographer to the Dutch East India Company from 1602 to 1619. He produced maps and globes that were used to record and promote Dutch trade to the East. He advocated trade around the Cape of Good Hope and to this end he worked on obtaining data on magnetic variation and star positions around the South Pole.

In the seventeenth century globe production developed in Italy. The Italians began by copying Dutch globes. In the latter half of the seventeenth century, globe production developed in France and England as globe production again followed the centers of trading dominance.

Globes were costly to produce. They involved highly skilled techniques of engraving and accurate cartography that required adept, expensive craftsmen. Most globes were produced by drawing maps on pieces of manuscript—called gores—then attaching them to a sphere. Martin Waldseemuller made the first printed globe gores around 1507. Some globes were painted directly onto the sphere, such as Johannes Schoner's 1520 globe. The very wealthy could commission more elaborate designs. Paolo Forlani made a globe etched on silver in around 1560.

The globe was a gridded world. The globes came with divisions of latitude and longitude, degrees and minutes, marks that held the world in a net of lines and intersections. Renaissance globes bear the imprint of the grid, which has now become part of our assumed view of the world. Contemporary television news programs around the world often reference the earth as a gridded sphere simply by presenting just a few lines of latitude and longitude in a spherical motif.

Globes were both educational and decorative. They were also used to encourage expedition and trade. In 1592–93 the English

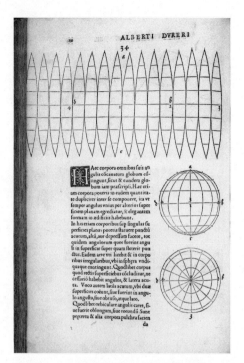

13. Diagram of globe gores in Albrecht Durer's *Albertus Durerus Nurembergensis pictor huius aetatis celeberrimus,* Paris, 1535. Courtesy of Dibner Library for the History of Science and Technology, Smithsonian Institution Libraries, Washington, D.C.

merchant William Sanderson commissioned a globe to be made by Emery Molyneux to encourage expedition and trade with China and India.

There was also a flourishing trade in books that taught people how to make and use globes. Peter Apian's *Cosmographia* (1524) contains material on using the globe. The frontispiece in a number of editions includes a large globe. (I will discuss Apian further in the next chapter.) The globe is also the central topic of Giovanni Paolo Gallucci's 1589 *Theatrum Mundi*. The first diagrams in the book are a pair of celestial and terrestrial globes. Detailed discussions of globes appear more frequently in the vernacular as the sixteenth century progresses, indicating that the use of globes is extending beyond the narrow range of Latin-reading scholars. In the second treatise of Robert Record's *The Castle of Knowledge*, published in 1556 in English, the master shows the scholar how to make a globe and an accurate armillary sphere. In the third treatise, which discusses the use of the sphere for "certain conclusions of daily appearances and other lyke matters," the master uses the sphere to explain the true location of places in the world.

The globe has entered the world as a symbolic as well as a material presence. When Renaissance princes and even contemporary presidents have their portrait made with a globe, an image of power and knowledge is invoked and represented. The globe appears as a symbol of power and leadership. While working at the Smithsonian Institution in Washington, D.C., on some of the works cited above, I visited the National Portrait Gallery. It struck me how many of the portraits included the ubiquitous globe, either as background or covered with a pointed finger: George Washington, Thomas Jefferson, Daniel Webster, Andrew Carnegie, Jedidiah Morse, and Simeon de Witt all featured globes in the paintings. The famous and the powerful—the not so famous and powerful—all bound by the same icon of the world. The globe is not simply a representation of the world; its depiction in portraiture invokes notions of power and symbols of authority.

"A Plane Surface"

In Book 1 of his *Geography*, Ptolemy outlines two map projections, different ways to represent the world on a plane surface. The translation of his work in Western Europe inaugurated an explosion of map projections, including simple conic (1506), cordiform (ca. 1500), oval (ca. 1510), octant (1514), gore (1507), Mercator (1511, 1569, and 1599), and sinusoidal (1570). The world was represented as an oval, a circle, a human heart, and was divided into eight and subdivided into a series of overlapping gores. Ptolemy's original projections were modified and transformed into multiple representations. New projections were also developed. In his 1514 Latin version of *Geography,* Johannes Werner (1468–1522) outlined the cordiform and oblique stereographic projections. Ptolemy's work provided a stimulus and a model for major developments in map projections.

There were also practical reasons behind some of the map projections. The portolans were prone to error. These charts used a simple projection that did not take into account the curvature of the earth, thus distorting the east-west direction at higher latitudes. This was not a major problem in the earliest years, when their use was restricted to the Mediterranean and the Black Sea. With the later voyages of discovery, however, ships began to sail in higher latitudes, where the error became magnified. The problem was solved by Mercator. In his world map of 1569, titled *New and Improved Description of the Lands of the World, amended and intended for use of Navigators,* Mercator used a map projection that now bears his name, in which the ratio between latitude and longitude remains constant. This involved increasing the value of latitude toward the poles. The map thus exaggerates the size of the polar regions but has the tremendous practical value of representing the world in a manner that allows a navigator to use a straight line to plot a course.

Extending the Grid: Cartography and Exploration

The dissemination of Ptolemy's work gave an extra stimulation to exploration. A Latin version of *Geography* had been available in Florence since 1406 and formed the basis for a series of symposia in the city between 1410 and 1440; these meetings involved a reassessment of knowledge of the known world and discussions of possible routes to the Far East and Africa. In the 1420s, Henry the Navigator's older brother Prince Pedro came to Florence to collect maps that may have prompted and aided Portuguese exploration along the coast of West Africa.

Ptolemy assumed the world to be spherical. The world covered 360 degrees, but he showed only the 180 degrees of what he believed was the habitable world. The Ulm edition, for example, stops just west of Europe and extends east only as far as East Asia. By showing only part of the world, Ptolemy left an enticing question. Renaissance scholars, princes, and explorers looked at Ptolemy's world map and wondered: What was on the other side?

Distance to the other side of the earth was also made less uncomfortably long by Ptolemy, who had measured the earth but underestimated the circumference by around 25 percent. He also overestimated the Eurasian land mass and underestimated the size of the oceans. It was with this picture of the world that Columbus undertook his momentous voyage. This is not to diminish Columbus's venture, but the Ptolemaic world was a slightly less-forbidding prospect.

Not only did Ptolemy's *Geography* stimulate exploration, but it also allowed new discoveries to be incorporated. Ptolemy's grid was infinitely extendable. The grid was a subtle device capable of bending the world to fit the new discoveries. In 1520, Schoner drew the world on a double hemispheric projection so that the European discoveries of the New World and the Orient could more easily be represented. In order to include the New World, which stretched the known world beyond Ptolemy's "habitable world," Matteo Contarini (d. 1507) doubled the meridians in Ptolemy's original, from 180 degrees to a full 360 degrees. Throughout the sixteenth century, European discoveries and new map projections went hand in hand.

In 1597, Cornelius Wytfliet (d. 1597), geographer and secretary to the Council of Brabant, published his *Descritionis Ptolemaicae augmentum.* This supplement to Ptolemy was the first atlas of America. It contained eighteen regional maps of America and a world map that clearly showed the "other" 180 degrees. It was no paradox that Ptolemy's name was invoked in the title of this first atlas of the New World.

3

ENCOMPASSING THE WORLD

In his book *Cosmographia,* first published in 1524, Peter Apian made a distinction between cosmography, geography, and chorography. Cosmography was the study of the universe, geography was the examination of the earth, and chorography looked at specific parts of the earth. The divisions were not demarcations into separate fields of inquiry; rather, they reflected a relative division of a unified view of the world. The division between the terms was fluid rather than fixed; Ptolemy's work was sometimes translated as *Geography* and other times as *Cosmography.* The divisions were important scale distinctions of global, regional, and local—a similar gaze but different lenses.

In Renaissance cosmography the sphere was both an important metaphor and the basic unit of order. The earth was visualized as being perfectly round, the embodiment of a divine geometry. Throughout the medieval period and early Renaissance it was assumed that the earth must be spherical because the sphere is the most beautiful shape of all bodies—perfect in symmetry, as befits God's creation. The sphere of the earth was also capable of rational measurement and mathematical plotting. It allowed a purposeful gridding, the laying down of the mesh of lines of latitude and longitude. Renaissance cosmography sought to map the universe: celestial and terrestrial.

Renaissance cosmography incorporated medieval notions of macrocosm and microcosms as well as ideas of astrology and prophecy; it was a rich stew of rational measurement, religious

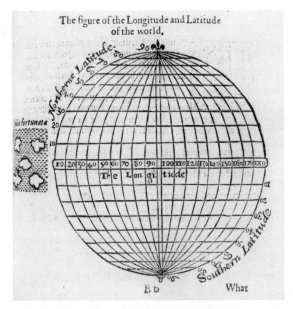

14. Sphere in Thomas Blundeville's *A plaine treatise of the first principles of cosmographie*, London, 1597. Courtesy of Dibner Library for the History of Science and Technology, Smithsonian Institution Libraries, Washington, D.C.

meditation, and esoteric discourses. Sixteenth-century cosmography marked no sharp epistemological break between the medieval and early modern worldviews. Even the technical gaze of the sixteenth-century cosmographers was not untouched by older conceptions. A geocentric worldview, a belief in astrology and the building blocks of the four elements (earth, air, fire, and water) mingled with new forms of measurement and survey. Cosmographical texts contained both new ways of seeing the universe as well as traditional conceptions of the world.

In this chapter I will look at a variety of cosmographers. I have categorized them roughly into three groups: the technical gaze of Peter Apian, Oronce Fine, and three English cosmographers, Robert Record, William Cuningham, and Thomas Blundeville; the cosmo-geographies of Sebastian Munster and Andre Thevet; and the cosmo-mysticism of John Dee. In reality, technical considerations, geographies, and mystical elements were bound up in the work of each cosmographer, but the demarcation allows me to highlight these different themes with less repetition.

Cosmographers were more than just theoreticians; they were also practitioners of earth measurement. Practical concerns with mapmaking, navigation, cartographic projections, and instrument-making and design were central elements in the whole project.

The term *cosmo-geography* deserves some comment. In the sixteenth century there was not a clear-cut distinction between cosmography and geography. The terms were used interchangeably in both book titles and official designations. Cosmography and geography coincided in the writing of world geographies; thus I will use the term *cosmo-geographers,* despite its cumbersome and somewhat ugly feel, to describe this intersection. Cosmo-geographers such as Sebastian Munster and Andre Thevet were writing geographies of the world. They were adapting the cosmographical project to write universal geographies at a time when the known world was rapidly increasing in size and complexity.

John Dee's distinctive position at the intersection of knowledge and power, science and the occult, deserves a special mention. While little of his written work remains, he was one of those pivotal figures who linked Elizabethan England with mainland Europe in shared ties of intellectual endeavor. He wrote, mapped, and plotted (in the double sense of the word)—his projects included encouraging English imperial expansion. His cosmo-geography covered a wide sweep of knowledge, some of which we would now call occult. He was the Renaissance magus who probably served as the model for Shakespeare's Prospero.

The threefold categorization that I have used is a useful starting point rather than a destination. There are no hard and fast distinctions. Apian and Fine cast astrologies and made maps as well as wrote on instrumentation; Dee made maps and brought back surveying instruments from mainland Europe to England as well as communicated with angels.

The Technical Gaze

Throughout the sixteenth century a number of influential books were published that outlined a cosmography. The single most influential work was by Peter Apian (1495–1552). Apian, also known as Peter Bienewitz or Bennewitz (his name was Latinized as Petrus Apianus), was born in Lesinig, Germany, on April 16, 1495. He studied at Leipzig and Vienna and was considered an outstanding mathematician, astronomer-astrologer, and instrument maker. He was appointed court mathematician to Charles V and made a Knight of the Holy Roman Empire. He held a professorship of mathematics at University of Ingolstadt, where he remained until his death in 1552.

The first edition of Apian's *Cosmographia* was published in 1524. It is a general introduction to a cosmography that included such subjects as astronomy, geography, surveying, navigation, and instrument making and design. It includes a variety of woodcut illustrations.

The work has a rather complicated publishing history.[1] It was first published in 1524. Five years later the cartographer-surveyor Gemma Frisius (1508–55) produced a corrected, amended version, published in Antwerp. In 1531 a shorter version of the initial book, with many woodcut illustrations, was published as *Cosmographiae Introductio* in Ingolstadt, Germany. Frisius produced revised copies, with new additions, including "Concerning the method of describing places" (1533 edition), "The use of astronomical rings" (1539 edition), and "The universal astrolabe" (1584 edition). The book was a huge success, especially the revised edition of 1533. Over an eighty-five-year period, the book was published in forty-seven separate editions (thirty-two in Latin, eight in Dutch, five in French, and two in Spanish) and in seven cities (Landshut, Ingolstadt, Antwerp, Venice, Cologne, Paris, and Amsterdam). It became *the* handbook of Renaissance cosmography.

A key illustration in Apian's cosmography, found in most of the various editions, is shown in illustration 15. It is a depiction of an armillary sphere, an instrument first outlined by Ptolemy in *Geography*. The armillary sphere is an astronomical model of our

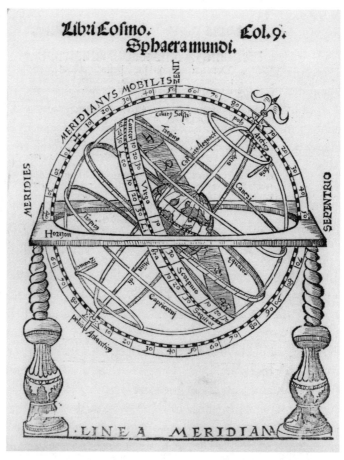

15. Diagram of armillary sphere in Peter Apian's *Cosmographicus liber*, Landshut, 1524. Courtesy of Dibner Library for the History of Science and Technology, Smithsonian Institution Libraries, Washington, D.C.

immediate universe that depicts the great circles of the heavens including the horizon, meridian, equator, tropics, polar circles, and an ecliptic hoop. It represents the world held in a cradle of measurements, surrounded by a celestial heavens. Everything is bound in a perfect sphere, capable of measurement and observation.

The book speaks to a terrestrial and celestial interaction. The earth is firmly placed in a wider universe. The tilt of the earth is used to explain the different climate zones, and the movements of the sun and moon are used to explain seasonal and diurnal variations in the length of the day.

The book is a mathematical rendering of the world. A central feature of the book is measurement: the correct way to measure latitude and longitude, to calculate times of sunrise and sunset around the world, to ascertain the correct height and distance of distant objects, and to establish the position of celestial bodies. The book is full of diagrams and illustrations that depict observers looking at the world through a variety of instruments in order to attain accurate measurements. The world is the object of a technical gaze. Observers see the world through the eyeholes of astrolabes, quadrants, and nocturnals.

The book is full of scientific instrumentation. Both Apian and Frisius were instrument makers: Apian designed a surveying quadrant and an armillary sphere; Frisius made maps and globes, managed an instrument-making workshop, developed designs for a cross staff and astronomical rings, and outlined the system of triangulated surveying.

Along with the many illustrations of scientific instruments, the book also contains volvelles. These are paper devices that allow various calculations: the lunar clock allowed the determination of the time at night; the horizon was a device that showed the relationship between the local horizon and the polar axis; the altitude sun dial was used to tell the time in different latitudes. These delightful devices, which were built into the book, allowed the reader to pull the little strings tied to the paper instruments. The volvelles allowed a more interactive readership of the book.

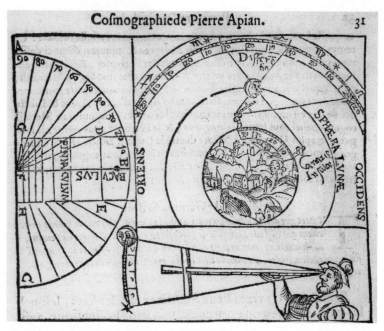

16. Diagram in Peter Apian's *Cosmographia,* Antwerp, 1581. Courtesy of Dibner Library for the History of Science and Technology, Smithsonian Institution Libraries, Washington, D.C.

The book is a compendium of devices that represent a developing technical gaze. But it would be incorrect to see it only as an example of the new science. Many of the diagrams draw directly from the earlier work of Pomponius Mela, Proclus, and especially Sacrobosco.[2] Like Renaissance cosmography in general, Apian's book contains practical devices as well as astrological assumptions, mathematical calculations as well as alchemical interpretations, empirical measurements as well as esoteric symbolism; it is a new way of looking that is deeply marked by older views of the world.

Apian was also a mapmaker. His first work, published in 1520, was a world map, *Typus Orbis Universalis,* based on the work of Martin Waldseemuller. His little pamphlet, *Isagoge* (1521–22), contained eleven propositions about the "diverse uses for this map."

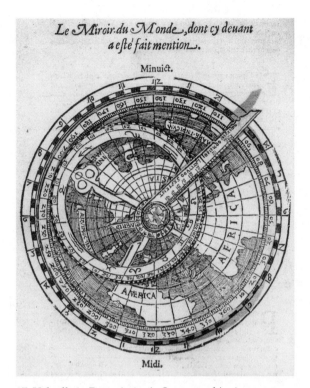

17. Volvelle in Peter Apian's *Cosmographia*, Antwerp, 1581. Courtesy of Dibner Library for the History of Science and Technology, Smithsonian Institution Libraries, Washington, D.C.

Various editions of the *Cosmographia* also contain a world map on a pseudocordiform projection.

Apian's work embodies the substance of Renaissance cosmography. He published a variety of maps and books, including an arithmetic manual for merchants, and wrote manuals on the use of various scientific instruments, including the armillary sphere, quadrant, Jacob's staff, and nocturnal. He described comets and wrote a major book on astronomy and astrology, *Astronomicum Caesareum* (1540). Astrology and astronomy were closely bound in Renaissance cosmography. It is reputed that Charles V gave Apian three thousand gold pieces for this work, one of the most lavish books of the sixteenth century. It is as much a scientific calculating machine as it is a book. Large, six-layered volvelles allowed complex, accurate calculations to within one degree. It is the book as computer.

Apian's cosmography reached a wide audience, including John Dee, Gemma Frisius, Sebastian Munster, and Oronce Fine.

Oronce Fine (1494–1555) was a mathematician, astronomer-astrologer, cartographer, and cosmographer. He was born near Briancon in France in 1494. His father and grandfather were physicians. He was sent to Paris in the first decade of the sixteenth century and took his master's degree in 1516. Fine was an illustrator and publisher who used many of his own woodcuts in the books he published. The title pages often bear his monogram and self-portrait. It was a recurring theme; he had a gift for self-promotion and is a constant presence in his books and maps. In 1531 Francis established a royal chair of mathematics. Fine had prodded the king by writing a poem that such a chair be established, and he was duly appointed to the position.

Fine's cosmographical works come in a variety of forms. First, he produced maps. His first independent work was his 1519 cordiform (heart-shaped) map of the world, dedicated to Francis I. His world map of 1531 is a double cordiform. It is an elegant cartographic projection, which Schoner used as a basis for his 1533 globe. Fine made his first map of France in 1525 based on plotting the latitude and longitude of 124 French cities, towns, and villages on a grid. His second major interest was in scientific instruments. He wrote on the principles and practices of making sundials; he published pamphlets on the design of the quadrant and devices for calculating longitude by lunar observation. Third, like many other cosmographers, he was fascinated by astrology. It is reputed that the unfavorable astrological prognostication he gave to the French prince landed him in jail. (Fine was jailed twice, once in 1518 and again in 1523). He wrote about the relationship between astrology and medicine. His interest in the esoteric side of the Renaissance project is evinced from his close relationship with the Elizabethan magus John Dee, who gave lectures at Paris in 1550 at the invitation of Fine. Dee had Fine's works in his library at Mortlake.

Oronce Fine published a number of cosmographical works and edited one version of Apian's *Cosmographiae Introductio*. His best-known work on cosmography is *De Mundi Sphaera*. It is a richly produced book, with exquisite woodcuts on fine paper. The title page shows Fine with an astrolabe, books, and surveying instruments. The book maps the cosmos, giving latitude and longitude of both stars and cities. While similar in basic outline to Apian's

18. Frontispiece of Oronce Fine's *Orontij Fine Delphinatis*, Paris, 1542. Courtesy of Dibner Library for the History of Science and Technology, Smithsonian Institution Libraries, Washington, D.C.

work—sharing the same concern with accurately mapping the world—there is less emphasis on scientific instrumentation and more emphasis on astrology.

A number of other works of cosmography were produced in the early sixteenth century, including *Cosmographicae aliquot descriptiones* (1537) by Johannes Stoeffler (1452–1531) and *Elementale cosmographicum* (1539) by Martin Borrhaus (1499–1564). Stoeffler's book is a compilation in Latin of many authors, mostly German. It includes a discussion of the geometry of a sphere with an explication of Ptolemy's two projections. Borrhaus's work is an illustrated handbook with sections on astronomy and geography. The 1551 version also contained a version of *Cosmographiae Introductio* edited by Oronce Fine.

Most of the cosmography books were illustrated, with the sphere being a recurring motif. In *Theatrum Mundi* (1589) by Giovanni Paolo Gallucci (1538–1621?), the sphere is the main object of analysis. The book has diagrams of celestial and terrestrial globes, along with a set of volvelles.

Apian's work was the most widely read of the cosmographers. His work was published in Latin as well as French, Spanish, and Dutch, but was unavailable in English. However, there were En-

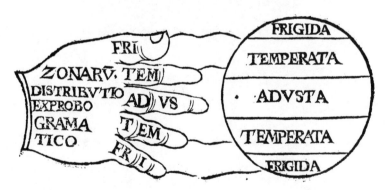

19. Diagram of climatic zones in Martin Borrhaus's *Elementale cosmographicum*, Paris, 1551. Courtesy of Dibner Library for the History of Science and Technology, Smithsonian Institution Libraries, Washington, D.C.

Encompassing the World ✣ 45

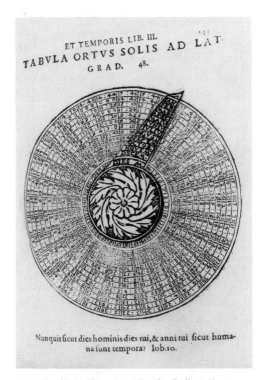

20. Volvelle in Giovanni Paolo Gallucci's *Theatrum mundi*, Venice, 1589. Courtesy of Dibner Library for the History of Science and Technology, Smithsonian Institution Libraries, Washington, D.C.

glish cosmographical works. I will examine three authors: Robert Record, William Cuningham, and Thomas Blundeville. All were influenced by Apian and Fine.

Robert Record's *Castle of Knowledge* was published in 1556.[3] The book is dedicated to Queen Mary and Cardinal Pole. The subtitle is "the explication of the sphere both celestiall and materiall." In the frontispiece divinity and knowledge are contrasted with ignorance and blind fate. Knowledge stands on a secure flat surface, while ignorance balances precariously on a ball. Knowledge, the image suggests, will release us from the whims of fate. The book is

21. Title page in Robert Record's *The Castle of Knowledge*, London, 1556. Courtesy of Dibner Library for the History of Science and Technology, Smithsonian Institution Libraries, Washington, D.C.

in the form of a Socratic dialogue between a master and a scholar. The scholar, when asked by the Master what he desires, replies, "I see in the heavun mervailous motions, and in the reste of the worlde strange transmutations, and therefore desire muche to know what the worlde is, and what are the principall partes of it, and also how all these strange sights do come."

The Castle of Knowledge is a very detailed, mathematically driven cosmography that demonstrates the ordered variations, mathematical relationships, and causal connections in the sphere. Record wrote clearly and succinctly, moving in a Socratic method from simple to complex terms. There is very little astrology. The zodiacal signs are discussed as simply divisions of the heavens. Record's cosmography is a forerunner of contemporary cosmology.

Throughout the sixteenth century cosmography was directed toward practical matters. William Cuningham's *The Cosmographical Glasse* (1559) was subtitled, "the pleasant principles of cosmographie, geographie, hydrographie or navigation." Cosmography is now being written for a more nonscholarly audience. We can compare Apian's and Fine's Latin texts, written for other scholars, with Record's and Cuningham's vernacular texts, intended for a more practically minded audience. In the preface Cuningham praises the use of cosmography as a practical art, which is especially useful in the defense of the country because it gives guidance in such practical matters as where to pitch tents and where to winter troops.

The book is in five parts. In the first, he uses a map of the earth and a map of the city of Norwich, his hometown, to make the distinction between cosmography, geography, and chorography. The second part is concerned with zones, parallels, climates, latitude, longitude, and eclipses. The third part covers mapmaking and surveying. Part four deals with hydrography and navigation, including tables of flood and ebb tides in the North Sea. The final part reads more like a straightforward geography book with very brief descriptions of places. America is assigned only two pages of terse comments, with the population summarized thus: "The people both the men and women are naked. Their weapons are bowes and arrows. They have warre with th'inhabitaunts of the country next." Cuningham was writing well before the sustained English imperial venture into North America.

In Cuningham's book the emphasis is on the practical arts of surveying, navigation, and geographic description. It is a handbook for national surveys and colonial expansion: an applied cosmography for the emerging state.

This gradual shift in cosmography toward practical, national purposes is also apparent in the work of Thomas Blundeville (fl. 1561). In 1594 he published *M. Blundeville his exercises*. The book is derivative of Record's *Castle of Knowledge*. Blundeville defined four parts to cosmography: "astronomie, the science of celestial bodies"; "astrologie, the science of the motions, aspects and influences of the stars both to foretell and prognosticate things to come"; "geographie," which described the whole earth; and chorography, which "described some part or region of the earth." This division of astronomy from astrology heralds the sharper division to come in the next and subsequent centuries. Blundeville's book went through a number of editions. In the sixth edition, printed in 1622, there is a new treatise on navigation, including the use of important navigational tools including the cross staff, mariner's cards, maps, and almanacs. Practical mapping and navigation were now figuring larger in the works, especially as English merchant ships were connecting a global trading empire with longer journeys across unfamiliar seas. Cosmography was being applied to practical concerns of commercial needs and imperial expansion, and in the process new separate areas of knowledge were emerging in their own right.

Geographical Concerns

Cosmography had a number of identifiable elements. As we have just seen, a technical concern expressed itself in early instrumentation. The sphere was encased in a grid that could be measured and understood. But a *narrative* grid was also placed over the world. The Renaissance project was interested both in technically understanding the world and in covering it in one narrative sweep. There were cosmographers—today we would call them geographers—who were writing a geography of the world.

The cosmo-geographers of the sixteenth century were part of a long and wide tradition of universal geographical coverage. Predating the Europeans of the sixteenth century were a number of Arabic cosmographers. Al Qazwini (1203–83), for example, wrote the illustrated book *Aja ib al Makhlaqat* (Wonders of Creation), which was part geography, history, astronomy, and natural history. It became a popular text and was translated into Persian and

Turkish. The first Hebrew book of geography, *Igeret Orhot Olam* (Epistle on the Itinerarium of the World) by Abraham Farrisol, first published in 1524, contained descriptions of the continents and the different climate regimes.

An important forerunner to the sixteenth-century works was the fusion of history and geography contained in the great chronicle of the world, printed in Nuremberg in 1493. The *Nuremberg Chronicle* was history and geography combined in a great rolling epic that went on for almost 600 folio pages. It was called by its creators *The book of Chronicles from the beginning of the World.* It is one of the great picture books. The illustrations are not merely small inserts—they are major pieces of the text. There were 1,809 prints from 645 woodcut blocks, including prints of 598 notable people, scenes from the Bible, maps, and city views. As benefits an enterprise that was in part a patriotic attempt to show the dominance of Nuremberg over Augsburg and Ulm, the view of the city of Nuremberg is the largest woodcut. It was a huge, ambitious undertaking to produce the biggest, most illustrated book to date. The book contained many arresting images, including some woodcuts by Albrecht Durer. The detailed figures, scenes, and cities come alive even to the modern reader—the images are more three-dimensional, more modern looking than was found in the flatness of traditional medieval art. It was first published in Latin in June 1493; a German edition came out six months later. Almost 1,500 copies in Latin and 1,000 in German were produced in the first print run. The book went through various editions in both Latin and German.

Although ostensibly a history of the world from the six days of creation to the date of publication in 1493, the book is also a geography of the world. History and geography were yet to be so divided as they are today. Maps of Europe are placed alongside elaborate genealogical trees. It is a partial geography of the world. The book did not mention India or China; apart from a brief mention of Egypt, Africa is also ignored. It presents us with a view of the world before the great discoveries. A Ptolemaic world map, for example, is included with no hint of the New

50 ✣ *Making Space*

22. View of Anglia in Hartmann Schedel's *Liber cronicarum*, Nuremberg, 1493; leaf 288r; woodcut. This item is reproduced by permission of The Huntington Library, San Marino, California.

World that had been invaded just one year earlier by Christopher Columbus.

Cosmo-geography makes an early appearance in the sixteenth century with the publication of Martin Waldseemuller's *Cosmographiae Introductio*, first printed in 1507. It was intended to accompany and explain a globe and world map. This small book contained the standard cosmographical model of a geometric world and introduced the reader to the sphere, axis, poles, circles of heaven, parallels, climates, and winds; as Waldseemuller noted, "a cosmographer ought to know especially the elevation of the pole, the zenith, and the climate of the earth." Waldseemuller also included the reports of the four voyages of Amerigo Vespucci. These purported to be letters from Vespucci claiming the discovery of America a year before Columbus. The letters were forgeries, probably made in Florence to allow the Florentine Vespucci to outshine the Genoan Columbus. The book's lingering legacy is that it conferred the name of America, in honor of Vespucci. The Florentine ruse was successful; in the ultimate in Italian city-state oneupmanship, the New World carries the name of a Florentine rather than the name of the Genoan who landed there in 1492.

Waldseemuller was an influential, though shadowy, figure in the history of cosmography. His birth and death dates are hazy approximations. Born in Hapsburg, Germany, around 1475, he attended Freiburg University. He moved to Saint-Die in the western Vosges in Lorraine around 1505 and lived there, with visits to Strasbourg, until he died around 1518–21. He became a canon of the local Benedictine monastery. Waldseemuller is best known for his maps and globes.[4] His large map of the world that accompanied the *Cosmographia* was copied by both Peter Apian and Sebastian Munster and formed the basis for Schoner's globes. It was the first printed map of the world to contain the New World. He also provided maps for the 1513 edition of Ptolemy's *Geography*, considered by many to be the first modern atlas because the new maps conform to a standard design. His 1516 *Carta Marina* is the oldest printed nautical chart.

A narrative grid was placed over the world in cosmographical texts, but it showed a world in flux. The cosmo-geographers were faced with a difficult task. The sixteenth century saw a huge expansion of geographical knowledge. New lands were added and new reports clashed with classical authority. The cosmo-geographers of the sixteenth century had a paradoxical sense of both an ordered world and a world that was subject to dispute and alternative interpretations—order and disorder, coherence and dissonance. Strategies to deal with this paradox varied from a belief in historical authority to a reliance on contemporary geographical writing. Sebastian Munster (1488–1552) seemed happy to accept traditional authority, while Andre Thevet (1516–92) seemed to rely on firsthand accounts. The distinction was in part due to their respective areas of inquiry: Munster dealt with the Old World while Thevet was trying to comprehend the New World, where classical authority was understandably absent. There was rarely a clear-cut distinction between the two approaches; rather, there was a shifting compromise. There was neither a rapid jettisoning of the ancient authority nor a wholesale embrace of new observations. The ancient texts and new worlds were interwoven, interleaved in a complex cosmo-geography.

Sebastian Munster was born near Mainz. He came from a religious family, both his grandfather and uncle having been priests. In 1505 he went to Heidelberg to enter a Franciscan order. Two years later he went to Louvain to study mathematics, geography, and astronomy. In 1509 the order sent him to the monastery of St. Katharina in Rufach in the Vosges mountains to study geography, mathematics, cosmology, and Hebrew. Hebrew scholarship and geographical understanding were recurring interests in Munster's professional life; all were part of understanding God. For him, there was no paradox between established religious faith and the new forms of knowledge.

Munster was a prolific writer, publishing eighty books on theological subjects and translating a Hebrew text of the Bible into Latin.[5] Munster continued his travels, extending his knowledge by visiting centers of intellectual activity. He stayed in Tübingen and in 1518 moved to Basel. In 1524 he was appointed professor of Hebrew at Heidelberg University, where he also lectured on mathematics and cosmography. In 1525 his broadside *The Instrument of the Suns* appealed to German scholars to send him maps and sketches of German lands so that he could make a detailed map of the country. In effect, Munster was making a national geography.

Like all cosmographers and mapmakers, he was in thrall to Ptolemy. His *Ptolemy's Geography* first came out in 1540. It went through numerous editions with added notes and new, woodcut maps. Munster used Ptolemy's projections and maps but also introduced new maps. There was a national bias; there were ten maps of different German regions. In Munster's *Ptolemy* the "original" maps of Ptolemy were included along with new maps. *Tabula Moderna* are placed side by side with Ptolemaic maps—the old and

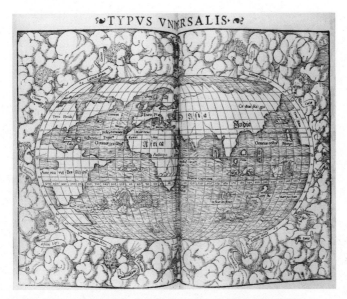

23. World map in Sebastian Munster's edition of *Ptolemy's Geography*, Basel, 1552. Courtesy of Dibner Library for the History of Science and Technology, Smithsonian Institution Libraries, Washington, D.C.

the new hurled together. The Renaissance cosmographers were at a crossroads where old and new maps were printed together as if the project was on a cusp, delicately balanced between a rediscovered old and an uncertain new. The maps, often side by side, with Ptolemy's older map placed beside the newer map, were an indication of the tension embodied in Renaissance cosmography between the reliance on classical authority and the commitment to contemporary scholarship.

Munster's major work, *Cosmography*, was first published in Geneva in 1544. It was a massive work. The first edition had 659 pages with 520 woodcut maps and illustrations. Subsequent editions increased in size. The 1548 edition enlarged the work to 818 pages and 725 woodcuts. By the 1550 version, the work had reached gargantuan proportions of 1233 pages and 910 woodcuts. It was published in all the major European languages as well as in

54 ❖ *Making Space*

24. Regional map in Sebastian Munster's edition of *Ptolemy's Geography*, Basel, 1552. Courtesy of Dibner Library for the History of Science and Technology, Smithsonian Institution Libraries, Washington, D.C.

Latin. It was amended and added to in the various language editions and shortened to form concise handbooks. The 1575 Paris edition, for example, was augmented by Francoise de Bellaforest. A slim handbook titled *A Briefe Collection and compendius extract of strange and memorable things gathered out of the Cosmography of Sebastian Munster* was also published in London in 1574. Thirty-six complete editions were published between 1544 and 1628. For much of the sixteenth century it was the single most important source of geographical, historical, and scientific knowledge. It was the great educational book of its era.

An eclectic collection of material, some old, some new, part old myth, part new fact, the book contains material on surveying and mining techniques as well as discussions of the phoenix, goblins, and spirits; it includes material detailing the reasons for sugar growing in parts of Italy as well as a discourse on the one-eyed and large-eared people who were supposed to inhabit parts of India. Discussions of latitude and longitude sit side by side with genealogies of long-dead European monarchies and Biblical exegesis.

Munster's work encapsulated both new knowledge and older conceptions of the world. Many of his maps have extensive border decorations: the one for Asia, for example, shows images of monsters, derived from Pliny, of men with no heads but eyes and mouth on their chest, and men with dogs' heads. These images came from the fourth century B.C.E., when India was described as a land of marvels populated by strange beasts that were half-human, half-animal, such as the *martikhora*, which had a man's face, a body of a lion, and the tail of a scorpion; and deformed people as the *sciapods*, who had a single giant foot. These tales were repeated down through the years. Although Munster was doubtful of their existence, their visual appeal were popular with readers. In a 1545 edition of *Cosmography,* he noted that nobody "has ever seen these marvels. But I will not interfere with the power of God, he is marvellous in his work and his wisdom is inexpressible."

Book 1 of *Cosmography* contains a basic cosmography. Book 2 is a geographical-historical description of provinces of Europe, with Germany comprising almost six hundred pages in the later editions. Subsequent books deal with the rest of the world. Munster's cosmography is biased toward Europe in general and Germany in particular. The further away from Europe, the more classical references to mythological people were cited. Far-flung places were more easily treated as exotic because they lacked the steady returning stream of more reliable and recent information. While Munster's cosmography may appear at first sight as a morass of material, it shows a firmer, more precise geography in the process of emerging. Munster was creating a new textual form, the geographical narrative. Munster's *Cosmography* marked the beginnings of a distinct geographical discourse of space and place. The book is a metaphor for the state of geographical knowledge itself:

expanding, rambling, slightly incoherent. But some order was beginning to be imposed on a world that was growing and expanding beyond traditional boundaries of epistemological understanding.

The success of Munster's book was due in no small part to the ambitious comprehensive coverage; the world was being ordered, comprehended in one narrative sweep, encompassed in a single text. In one way, it is easy to see Munster as a prisoner of classical errors and medieval prejudices, his cosmography still locked in the past. Yet by bringing them together, by encompassing the world in all its (mistaken and real) variety, Munster was creating the basis for a more accurate universal geography. Cosmography was living up to its promise of encompassing the world.

A large part of the readership was attracted to the tales of sundry customs and strange rites. There was a popular fascination with the exotic. Munster's book was a prototype of the early *National Geographic* model, full of bizarre ways and salacious details. Munster's readers were no doubt both attracted and repulsed by the stories of one-legged giants and cannibals in faraway lands. Geographical narratives have always been more popular the more they encompass the bizarre and the exotic.

More accurate knowledge of distant lands could only be ascertained by exploration and discovery. The writings of the many intrepid travelers filtered through into the cosmographical writings. Andre Thevet traveled to exotic lands and wrote about them. He was educated at Poitiers and then Paris, where he secured an appointment with the Cardinal of Amboise. He developed an interest in travel and went to the Middle East in 1549 under the patronage of the Cardinal of Lorraine; he remained there for four years. He published his *Cosmographie de Levant* in Lyon in 1554. The next year he was the chaplain to Villegagnon's Huguenot settlement in Brazil, but he became ill and soon returned to France. Later, he claimed to have taken a return route through Cuba, Florida, and Canada.

Thevet published *Les Singularitez de la France antarctique* in Paris in 1557 to a very favorable reception. With this book, we witness the incorporation of America into cosmographical understanding.

25. Illustration in Andre Thevet's *Les Singularitez de la France antarctique*, Paris, 1557. Courtesy of Newbery Library, Chicago.

It is a compendium of custom, religion, and natural history, as well as a basic geography. It is one of the first French books about the New World. An English version was published in 1568 titled "The New Founde world: wherein contained wonderful and strange things." The subtitle proudly notes that the work will reform "the errours of the ancient Cosmographers." Precision and accuracy are assured, but the subtitle also promises to reveal many wonderful things. Like the modern-day makers of horror movie parodies that both deliver and mock the horror scenes, some sixteenth-century cosmographers wanted to correct the ancient cosmographers but also use some of their salacious imagery.

Thevet subsequently obtained a court appointment and was made Royal Cosmographer to four French kings. *La Cosmographie universelle d'Andre Thevet, cosmographe du Roy* appeared in 1575. Despite its title, it is mainly about the New World and North America in particular. In the book Thevet claimed to have spent

26. Illustration in Andre Thevet's *La Cosmographie universelle d'Andre Thevet, cosmographe du Roy*, Paris, 1575. Courtesy of Newbery Library, Chicago.

twenty days in Canada talking with the natives. Much of the description is written as if Thevet spent considerable time in North America. The book was not well received. It was widely criticized by both Protestants and Catholics, a rare achievement in such divided times. An English critic referred to the work of Thevet and Bellaforest (a rival cosmographer and arch-critic of Thevet who published a cosmography in the same year) as "wearie volumes . . . untruly and unprofitablie amassed and hurled together."[6]

Thevet borrowed from a variety of sources, including his countryman Jacques Cartier (1491–1557) and especially from Ramusio's *Delle navigationi et viaggi*, published in three volumes in 1550, 1556, and 1559 in Venice with maps of the New World and New France. A contemporary referred to Thevet's work as "second-hand rags and tatters." His accuracy and authenticity are open to question. He confused deities from Canada in his discussion of religions in Mexico. His use of other people's work as his own and his exaggeration of his own travels lead to a part-reported, part-fabricated geography. Thevet's geography is not so much discovered as invented.[7]

There are differences between Thevet and Munster. Thevet is trying to incorporate the New World into the cosmographical project while Munster is biased toward Europe and Germany. Thevet's work reflects the French expansionist drive to the New World; Munster's is a central European cosmography. There is also a difference in method and approach. Munster relies upon scholarship and written reports while Thevet makes a claim to witness. While Munster draws upon the tradition of scholarship, Thevet introduces the notion of personal experience. Munster repeats the classical authors while Thevet prefigures a more modern point of view. Munster relies on texts, while Thevet's distinctive fictionalizing is on the cultural brink of a more modernist subjectivity.

But there are also similarities. In both Munster and Thevet we can see the emergence of a distinct geography from cosmography. It is not a geography untouched by the past. Older myths persist, and even contemporary reports are not completely reliable guides. The explorers and discoverers came back with stories, but they are stories with an agenda. It is not fact replacing fancy, but one myth replacing another. Even so-called firsthand observers could not be relied upon. Travel writing and geographical writing were always subject to bias and distortion. But a geographical discourse is developing. It is a geography with a wider arc than we have today: it includes a plethora of observations, genealogy, and speculation as well as geography and reportage. If the cosmo-geographies were a form of fiction, they were becoming a more consistent fiction.

John Dee

The cosmographers of the sixteenth century were a varied bunch: Catholic clerics and militant Protestants, retiring scholars and people with an eye to the main chance, careful examiners of the texts and sloppy compilers. But they shared a number of general characteristics; their knowledge included astronomy and astrology, geography and chorology, instrumentation and mapmaking, the practical arts and artisan skills as well as scholarly endeavors and mystical deliberations. John Dee (1527–1608) was one of the most interesting cosmo-geographers of the sixteenth century.[8] Dee read widely and drew upon a large body of knowledge, some of which we would now call occult. He was also concerned in affairs of state; he gave practical help to the imperial expansion of Elizabethan England. In a letter to the Archbishop of Canterbury probably written in 1592, John Dee wrote, "I have lived, and still live, good, lawfull, honest, christian and divinely prescribed means to attaine the knowledge of those truthes . . . for [God's] honor and glory and for the benefit and commoditie publique of this kingdome."

Scholar and public servant, savant and secret agent, Dee is an exotic character whose life and interests capture the zeitgeist of his age.[9] He has not left a wide body of work. There is no cosmography that bears his name like Munster's or Thevet's, yet he had an enormous influence. He was part of a wider intellectual movement of Renaissance learning and Elizabethan statecraft. He had a huge library of cosmographical texts.[10] His library consisted of more than 2,500 books at a time when the entire holdings of the University of Cambridge probably did not exceed 400 books.

Dee was born in Wales. His father was well connected at the court of Henry VIII and sent him to Cambridge in 1542. Five years later Dee went to Louvain and studied with Frisius, who was then Cosmographer to the Emperor as well as the reviser of Apian's *Cosmographia*. Dee brought back scientific instruments made by Frisius, including the cross staff and globes as well as knowledge of the triangulation method of surveying. From Louvain, Dee moved to Paris, where he came into contact with Oronce Fine. Dee traveled widely throughout Europe, reaching Italy, Austria, Bohemia, and Poland, among other places. He returned to London in 1551 having made contact with an impressive network of leading cosmographers across Europe. He made frequent trips to Europe and corresponded with the two great map—and atlas makers, Abraham Ortelius and Gerardus Mercator. He was quite friendly with Pedro Nuñez, who was the Cosmographer Royal in Portugal. The two men discussed mathematics and navigation. When Dee was under detention for heresy he made Nunez his literary executor.

Dee was interested in mathematics. In 1570 he wrote prefaces to an English translation of Euclid's *Mathematics* and Record's *Ground of Artes*. Mathematics for Dee was a practical system of knowledge that could be used to survey the world and navigate the oceans. One of his students was Leonard Digges, who later wrote influential works on surveying and measurement. Dee invented a compass that allowed mariners to follow a route across a great circle. He was actively involved in exploration and discovery; he gave technical advice to various exploration knights, including Francis Drake, Martin Frobisher, Humphrey Gilbert, and Walter Ralegh, and instructed many seamen in navigational techniques. His students included Thomas Harriot and both Hakluyts, who, in turn, did much to promote English overseas exploration and discovery.

But mathematics for Dee was also a system of knowledge that allowed a glimpse of the divine. Dee was part of the magic-hermetic tradition, which included such people as Cornelius Agrippa (1486–1535), Giordano Bruno (1548–1600), Marsilio Ficino (1433–99), and Giovanni Pico della Mirandola (1463–94). Occult and mystical ideas and practices played a central role in the Renaissance.[11] Numbers were not just reflections of external objects; they were a mystical language through which the divine could be revealed and contemplated. In his 1577 text, *General and Rare Memorials,* he wrote of "one mystical city universall ... to meditate of the Cosmographicall Government thereof ruled under the King Almighty."

Dee was a magus and mystic but also sixteenth-century England's most practiced and theoretical scientist. Magic, science, and religion were part of one universal vision. It is difficult for us to imagine the seamlessness of science and magic, but in the sixteenth century the occult was part of the general system of learning and knowledge production. The occult practices included alchemy, astrology, and geomancy. Because Dee was deeply involved in all of these areas but left little writing, it may be revealing to mention the work of other scholars in these three areas.

Alchemy was a practical matter, an early form of chemistry, a science described by the influential work *The Mirror of Alchemy* by Roger Bacon (ca. 1214–94) as "transforming metals into another creating the elixir which fully perfects metals and imperfect metals." It was also a mystical art; the search to turn base metal into gold was also a search for the purity of the soul from the dross of mundane life. Drawing upon Arabic science and medicine along with Christian mysticism, alchemy was a system of knowledge concerned with transforming the world, improving health, and reaching a higher state of consciousness. Its principal elements were an ancient esoteric tradition of probing and changing matter, a sixteenth-century hermetic mysticism of revealing human nature, and a physician's chemistry of healing powers. Dee was involved in alchemical experiments at the court of Emperor Rudolf in the late 1580s.

Astrology books figured largely in early printing. European writers drew upon Arabic writings, which in part were translations and amendments to one of Ptolemy's other great works, his text on astrology, *Tetrabiblos*. Arabic writers such as al Qabrisi, Abu Ma'shar, and Ibn Abi al-Rijal were translated into Latin in the late fifteenth and early sixteenth centuries; Latin translations of these cited authors were published respectively in 1485, 1515, and 1485. Many of the cosmographers, including Apian, Blundeville, Fine, and Schoner, constructed horoscopes and astrological diagrams. Fine's *Les Canons et documens tresamples,* first published in 1543, is a small pocket book that includes sections on the signification of planets in the different astrological signs, auspicious times to take medicine, and the necessity of being aware of planetary alignments for "best beginnings and perfection of all works and human negogiations." Dee was Queen Elizabeth's astrologer and was

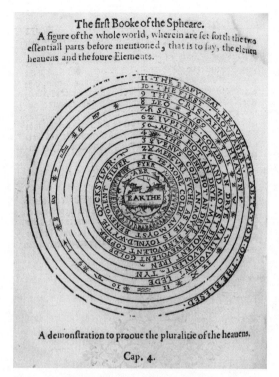

27. Illustration in Thomas Blundeville's *A plaine treatise of the first principles of cosmographie*, London, 1597. Courtesy of Dibner Library for the History of Science and Technology, Smithsonian Institution Libraries, Washington, D.C.

asked to find the most propitious date for her coronation; he decided upon January 15, 1559.

While many of us know what astrology is, geomancy is now less well known. Geomancy is the use of numbers, points, lines, and figures to help in foretelling the future. A number of geomancy works appeared in the sixteenth century. In 1574, Jean de La Taille (1533?–1611) published *La Geomance,* in which he defined astrology as being concerned with the celestial and geomancy with the terrestrial. In 1591, Francis Sparry produced an English

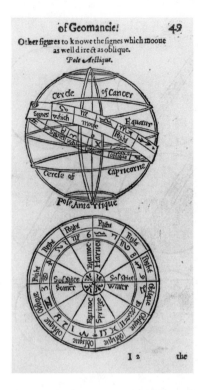

28. Illustration in *The Geomancie of Maister Christopher Cattan*, London, 1591. Courtesy of Dibner Library for the History of Science and Technology, Smithsonian Institution Libraries, Washington, D.C.

version of a French work, titled *The Geomancie of Maister Christopher Cattan*. It was subtitled a "wittie invention to knowe all thinges, past, present, and to come." Sparry/Cattan provided an alphabet with numerical values: A was 10, B was 2, C was 22, and Z was 14. This key was then used to ask a whole series of questions. On page 239, two examples are given. To know whether a person is telling a truthful tale, write their name and the name of the day, add up all the numeric values using the alphabet code, add 26, divide by 7, and if the remainder is even then the person told you a lie. To know whether the husband or wife shall die first, write the names of both husband and wife in Latin, add up all the numeric values and divide by 7; if the remainder is even the woman would die first, if uneven the man would die first. Dee was immersed in such prophetic works.

In the sixteenth century, science, as we know it today, had yet to emerge from a richer, wider tradition. Thus a writer like Leonardo Fioravanti (1518–88) could write about what we would now call alchemy, medicine, and science in the same text. His 1571 *Del compendio de i secreti rationali* is part science, chemistry, botany, and medicine, as well alchemy and magic. It is proto-science and folk knowledge.

Some of the rich mix of Renaissance magic can be seen in the work of Giovanni Battista Porta (1535–1615). Born in Naples, he was largely self-taught. He worked on optics and mathematics and, like Dee, he worked on cryptography. His book of magic, *Natural Magick* (to give it its English title), was published in 1558. Porta assumed a rational and ordered universe, which the magician had access to through study. His book made no distinction between natural magic and natural sciences. He included chapters on changing metals, beautifying women, tempering steel, invisible writing, and pneumatic experiments, which would not look out of place in contemporary handbooks, as well as advice on how to get women to cast off their clothes and go naked. He reported that, to accomplish this, conjurers write characters on a lump of hare's fat and mumble some words as they burn the fat. Porta writes that you do not need the mumbo-jumbo words—just burning the hare's fat will do. His book is a wonderful example of a half-distilled knowledge emerging from both experimental knowledge and the work of confidence tricksters.

From 1560 to 1583 Dee was at the very center of such Renaissance learning in England. He tutored and advised some of the most important and powerful people in the land, including Sir Walter Ralegh; the Duke of Northumberland; and the earls of Arundel, Pembroke, Lincoln, and Bedford. He was an important advisor to the queen. His mathematical and astrological knowledge was used in the revision of the calendar and in casting horoscopes to find the most propitious days for events and actions. He was also an Elizabethan secret agent, referred to by the queen as "My noble Intelligencer."

Dee was a pivotal figure in the early English attempts at overseas expansion and exploration. He promoted the idea of a British empire and was a technical advisor to many expeditions and explorations. He drew upon the broad range of European cosmographical works of Frisius, Fine, and Pedro Nuñez, as well as the geographical information of Mercator and Ortelius, to promote overseas exploration and expansion. He was in contact with other proponents of empire such as Sebastian Cabot and Richard Eden. He made maps of the northeast passage to China and the Far East. He was even granted all the land north of the 50th parallel in Canada by Sir Humphrey Davis in 1580. His 1577 *The Great Volume of Famous and Rich Discoveries* argued the case for English expansion into the Far East with notes on a possible northeast Cathay route. In the same year his *General and rare memorials pertayning to the Perfect art of Navigation* proposed an expansion of the navy to promote overseas development and the extension of the "Imperial Monarchy." His pupils, Thomas Digges and Thomas Harriot, played a major role in the development of mathematics and surveying in England.

The Narrowing Arc

The cosmographers of the sixteenth century encompassed a wide arc of human knowledge. The field of cosmography would later splinter into astronomy and geography, while astrology would join the nonscience category along with alchemy and natural magic. It would become more and more difficult to encompass the world. The scientific revolution involved a deepening, but also a narrowing, of what was considered knowledge. The sixteenth-century cosmographers were almost the last hurrah of an ambitious project. There were counter trends. Robert Fludd (1574–1637) carried on the Dee tradition. He had an ambitious plan for a great work that pulled together the interconnections between the macrocosm and microcosm. He wrote on magnetism, music, and alchemy, and was also interested in prophecy, geomancy, and astrology. His work was printed by Theodor de Bry as sumptuous picture books full of alchemical symbols and arcane designs.[12]

Fludd is the exception. By the eighteenth century the semiofficial title in Britain was no longer Cosmographer but Geographer to the King. Arguably, the last cosmographer in the old style is Alexander von Humboldt (1769–1859). Although he is outside our historical terms of reference, a brief summary of Humboldt's interests provides a fitting postscript to our discussion of cosmography and cosmographers. From 1782 to 1792 he studied at universities in Frankfurt, Göttingen, Hamburg, and Freiberg. Widely read, Humboldt had interests in economics, geology, botany, and mining. He arrived in Paris shortly before the storming of the Bastille and, as he wrote later, the event "stirred his soul." He had a lifelong social concern and radical political beliefs. In 1792 he joined the Prussian mining service, where he devised better safety lamps and rescue apparatus. But there was also a mystical side; he always sought to identify the "life force" (vis vitalis). Like Dee and Fludd, Humboldt was interested in mathematics and believed in universal harmony. His life's work was *Cosmos,* published between 1845 and 1862. He originally thought of calling it *Cosmography.* It was a popular scientific book subtitled *Sketch of a Physical description of the Universe.* He offered it, as he wrote, in the "late evening of a varied and active life" in order to "discern physical phenomena in their widest mutual connection, and to comprehend Nature as a whole, animated and moved by inward forces." By the time he turns to covering recent scientific theories in volumes 3 and 4, the book is unwieldy, weighed down with references. The unity is collapsing under the weight of the material. *Cosmos* was one the last great attempts to encompass the world. While Humboldt may be one of the first modern geographers, he was also one of the last cosmographers.

4

MAPPING THE WORLD

From the first printing of Ptolemy's *Geography* in 1475 to the end of the sixteenth century, the world was mapped in broad outline and some places were mapped in great detail. A rich variety of maps were produced: maps of estates, towns and cities, counties and principalities, countries and continents. City, national, and world atlases were created. Not only were maps made, but a culture of cartographic literacy was also created as maps were more widely read. Maps became important and influential texts. The mappings of the sixteenth century bridge the gap between the medieval and the modern cartographic representations of the world; they are the hinge on which we swing from the old to the modern, from then to now.

The mappings of the world were closely tied to the painting of the world and to the writing of the world; they were part of wider aesthetic and discursive representations. There was a strong connection between cartography and art in the Renaissance. So strong was this link, in fact, that there was little distinction between the two. Cartography was an integral part of the Renaissance revisioning of the world. Renaissance artists worked in a variety of media including architecture, painting, and sculpture as well as cartography. Albrecht Durer (1471–1528) drew world maps and wrote instructions for the drawing of globe gores. His use of the grid in painting was equally suitable to surveying and mapmaking. The Florentine Leon Battista Alberti (1404–72) was intrigued with applying perspective to both cartography and painting. Al-

Mapping the World ❦ 69

29. The grid and art. Illustration in Albrecht Durer's *Albertus Durerus Nurembergensis pictor huius aetatis celeberrimus*, Paris, 1535. Courtesy of Dibner Library for the History of Science and Technology, Smithsonian Institution Libraries, Washington, D.C.

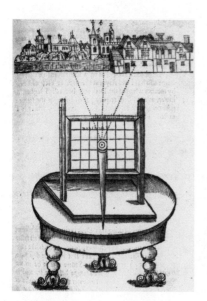

30. The grid and cartography. Illustration in Albrecht Durer's *Albertus Durerus Nurembergensis pictor huius aetatis celeberrimus*, Paris, 1535. Courtesy of Dibner Library for the History of Science and Technology, Smithsonian Institution Libraries, Washington, D.C.

berti introduced the polar or conic projection in his *Descriptio Urbis Romae*. There is a case to be made that the Ptolemaic grid was a central feature of the development of Renaissance perspective. The theory and practice of perspective—essentially tapering grids projected onto space—emerged at the same time as the diffusion of Ptolemy's *Geography*. Leonardo da Vinci, for example, was not only a great artist but also a skilled mapmaker. He drew bird's-eye views of the Arno River and plan view map of the city of Imola. He had a copy of Ptolemy's *Geography*, and his notebooks contain numerous mentions of the work as well as a more formal discussion of Ptolemy in regard to Leonardo's own plan to reveal the human body. Ptolemy's mapping project had given Leonardo an idea to organize his representation of the human body.

The distinction between art and cartographic representation was even more blurred when we consider the use of maps in decoration. During the sixteenth century it became fashionable to use maps as wall decorations. The Marquis of Mantua (1484–1579) commissioned maps and city views as decoration in his two country palaces of Gonzaga and Marmirolo as well as in his townhouse in Mantua. Cartographic images also appear in the media of fresco and mosaic. Cosimo de Medici (1519–74) employed two mapmakers to paint Ptolemaic world maps on the doors of his curiosities room in the Palazzo Vecchio. In the Caprarola Palace, built between 1559 and 1575 just north of Rome for the Farnese family, maps of the world, the continents, Palestine, and Italy covered the walls of the main reception room. Around 1580, Pope Gregory XIII commissioned a number of cartographic decorations for the Vatican. In the Third Loggia there is a huge (seventeen feet in diameter) map of the western hemisphere, designed by Egnazio Danti, a cosmographer and Dominican monk. Danti was also commissioned by the pope to decorate the Galleria del Belvedere with forty map frescoes of Italy.

Sixteenth-century maps were consciously produced as works of art. In their form and look they were pictorial representations of the world. The maps employed mimetic devices to convey the relationship between things in the world and things in the map. Little mounds represented hills and mountains, clusters of buildings located towns and cities, tiny trees indicated forests. Maps were pictures that became an important part of the visioning of the world.

The world was represented on maps in a simple, uncomplicated way. Little trees do look like a forest, and it is easy to read relief from the size and spacing of little molehills. But there were other layers of meaning. The maps represented not only the inert physical world but also complex social and political relationships. Maps were embedded in the denser messages of power and authority; they were written and read with sophisticated social nuances. Part art, part cartography, part political message, the maps of the sixteenth century were both very simple and very complex.

The connection between cartography and the book became more apparent in the sixteenth century. The development of printing and illustration techniques allowed cartographic images to be represented in texts; the medieval manuscript map became the modern printed map encased in a wider text. Manuscript maps were still made, often designed for specific purposes and particular audiences, but a wider cartographic literacy was created through the diffusion of printed maps. Maps became part of the printed word with the development of the printed atlas, or book of maps. The book and the map took on new relationships.

Renaissance mappings come in a number of forms and a variety of scales for a multiplicity of purposes. In this chapter I will restrict my comments to books of maps of the world, the nation, and the city. In an incredibly brief burst of cartographic ingenuity and innovation, arguably the first modern atlas (1570), the first city atlas (1572), and the first national atlas (1579) were all produced. Within the space of less than ten years, the standard representations of the world that we still use today were first developed. Modern space was contextualized in the Europe of the 1570s. It was a singular historical moment with a continuing legacy.

The Atlas of the World

One of the signal achievements of Renaissance cartography was the world atlas. Inside its covers, the world was gridded and plotted, mapped and surveyed. The atlas encompassed the world in one volume: the world as text, the text as the world.

The two most influential atlas makers of the sixteenth century—arguably of any century—were Abraham Ortelius and Gerardus Mercator. Abraham Ortelius (1527–98) is the author of one of the world's first atlases, *Theatrum Orbis Terrarum*. There had been collections of maps brought together in one volume before Ortelius, but he was to set a standard by which subsequent collections would be judged and compared.[1]

Ortelius was born, raised, lived, and died in Antwerp, a thriving, dynamic, merchant city. He always referred to himself as a citizen of Antwerp. In 1560 the city had a population of around one hundred thousand, including almost six hundred foreign merchants. The city was a crucible of European merchant capitalism, full of rich merchants and disposable incomes. The Portuguese had established their spice market in the city in 1499; Germans traded in metals; and it was an important center for English cloth merchants. The city exported a wide range of goods including furniture, jewelry, glassware, paper, maps, and musical instruments. It was also a major center for printing and graphic art. Ortelius was involved in the map trade. In 1547 he was listed in municipal records as a "painter of maps." He traded in maps, as well as antiquities, books, and coin, and around 1564 he started to make his own maps.

The *Theatrum* was begun around 1566–67 and by 1568 it was in full production. Most of all the engravings were done by Frans Hogenberg, who employed nephews of Gemma Frisius as his assistants. The volume was completed in 1569 and sold to the public in 1570. We know something about the making of the book, or at least one version of what happened, because five years after Ortelius died, Jan Radermaker, a lifelong friend of Ortelius, wrote his recollection of the genesis of the *Theatrum* in three letters to Ortelius's nephew Jacob Cool. Most maps, especially the larger maps, were single sheets, which were kept rolled up and unrolled only for examination. Radermaker worked for Gilles Hooftman, a rich merchant and shipowner who collected charts and maps. Hooftman wanted to get around the problem of having to roll and unroll maps every time anyone wanted to look at them. Radermaker passed the assignment on to Ortelius. The solution to the problem was to result in the *Theatrum*. This version of events gave high prominence to Radermaker. Other versions suggested that Ortelius had already collected maps from his travels and, as an active dealer in geographical literature and maps, thought about bringing them together in one volume for sale.

But Ortelius was not concerned merely with providing maps in the handier book form. He wanted the very best maps. He sought out the most accurate current maps and collected maps from a wide range of cartographers. In a gracious act, he listed all the names of the cartographers whose work he had used: 87 names

31. Map of the world in Abraham Ortelius's *Theatrum Orbis Terrarum*, Antwerp, 1570. Courtesy of the Library of Congress, Washington, D.C.

were mentioned for the first edition in 1570, 170 by 1595, and 182 in the 1603 edition. The world map alone cited 44 authors, while 20 authors were called upon for the map of America. Ortelius was generous with his praise. He described Mercator as the "prince of modern geography."

The title of the book comes from the Roman notion of the world as a disk, *orbis terrarum*, while *theatrum* implies a scene where action takes place, the setting of human life. Beginning around 1550 and in use until 1700, the word *theater* was used by many authors. It had multiple meanings: the world as a stage where people are actors and fortune is the stage director; nature as a theater with humans as spectators; a view of divine providence, a setting where God displays his skills; a title that announces encyclopedic intentions of surveying all of nature to provide complete and ordered coverage of a chosen topic.[2] After Ortelius and especially in cartographic works, its usage was generally taken to mean an exhaustive collection of material drawn from a broad array of sources. The success of Ortelius's work meant that the use of the term *theater* spread quickly throughout the European book world. Shakespeare in his 1599 *As You Like It* repeated the Ortelian image: "This wide and universall Theater/All the world's a stage."

The frontispiece of the *Theatrum* has elaborate imagery of maidens as the continents; Europe is at the top, holding a scepter as symbol of authority and rudder and steering the affairs of the world; a large cross signals the Christian religion. Asia is an oriental princess with gems and precious stones; she holds in her left hand incense, the fumes of which represent oriental mystery; Africa is a black maiden, her head glowing with the heat of the tropics. At the foot is America as an Amazonian warrior; beside her a severed head, a symbol of cannibalism, and a bust indicating unexplored lands. In the introduction Ortelius noted, "there are many that are much delighted with Geography or Chorography, and especially with Mappes." He reassured his readers that he had chosen the "best and most exact" maps.

The book was hugely successful. Between 1570 and 1598, 2,200 copies were sold. It was first published in Latin and later in Dutch,

32. Frontispiece of Abraham Ortelius's *Theatrum Orbis Terrarum*, Antwerp, 1570. Courtesy of the Library of Congress, Washington, D.C.

German, French, Spanish, Italian, and English. Further materials were added in later editions. The 1570 edition had 50 sheets, while the Italian edition of 1612 had 129. The book was printed as late as 1724, and almost 7,300 copies in eighty-nine editions were eventually printed. Amazingly, 2,000 copies still exist.

The book's success owed as much to Ortelius's canny business practices as it did to the quality of the product. Ortelius bought his own paper, supplied copper engravings at his own expense, and sold directly to both customers and booksellers. Ortelius's *Theatrum* was a customer-oriented, quality-driven business venture. He had a keen business eye. The man from the merchant city produced a well-marketed book of maps. The financial success of the *Theatrum* provided a good living for Ortelius. He lived a burgher's happy life moving into ever larger, more expensive houses.

The book was widely respected: Philip II kept a copy close by him and appointed Ortelius "His Majesty's Royal Cosmographer" on May 20, 1573, giving him a golden necklace worth one thousand ducats. Ortelius would describe himself as "citizen of Antwerp and Geographer to Philip II of Spain."

The first edition is a collection of fifty maps. There is little writing. Ortelius's book is different from the loose and baggy monster that was Munster's *Cosmography*. The *Theatrum* is a more modern atlas, perhaps the very first modern atlas. Ortelius wrote in his introduction that geography was the eye of history; "I omit the reading of Histories," he wrote, because, as he went on to note in a stunning break with tradition, it is more pleasant when "the mappe being layed before our eyes, we may behold things done or places where they were done, as if they were at the present and doing." The map was replacing the text and geography was breaking away from history. While Munster's cosmography is more akin to the *Nuremberg Chronicles,* Ortelius's *Theatrum* is closer to a modern atlas.

Most maps in the *Theatrum* are gridded and plotted. In some cases, as with the world map, the grid lines run through the map; in other cases coordinates are given in the border. Ortelius's map of the world is followed by maps of the New World, Asia, Africa, and Europe. Larger-scale maps then follow. There are no detailed maps of the New World. The maps use pictorial icons to represent towns, mountains, and forests. There is little decoration in the map margins, although there are occasional ships and whales in the oceans. There is a compass for scale and sometimes a cartouche; but overall a quiet elegance defines the look and feel of the individual maps as well as the whole book. There are no "Ptolemaic maps"—only modern maps to represent a modern world.

The short bits of writing that accompany the maps are less a standard historic-geography and more of a textual discussion of the map itself, citing both ancient and modern authors. The discussion of the world map, for example, cites Ptolemy and nineteen other ancient authorities as well as forty-one more recent authors including Apian, Munster, and Thevet. Similarly, the map of the

33. Map of Iberia in Abraham Ortelius's *Theatrum Orbis Terrarum*, Antwerp, 1570. Courtesy of the Library of Congress, Washington, D.C.

New World draws on the work of Spanish writers such Coronado, Cortez, and Nunez, as well as French authors such as Cartier and Thevet.[3] Ortelius's atlas is a history of geographic thought. Ortelius presents the map as a text with multiple authors rather than a statement of fact or simply as filler for a rambling chronology.

Ortelius's atlas marks a revolutionary change. It embodies the power of the grid and shifts the description of the world: geography replaces chronology, space replaces time. The world is located less in chronological terms and more by spatial coordinates. Perspective rather than narrative dominates. It is the book of the geographical gaze, the text of the chorographical eye.

Mercator's Atlas

The man who gave us the term *atlas* for a collection of maps was Gerardus Mercator. The maps in his atlas were published over a number of years. In 1585 he published volumes covering France,

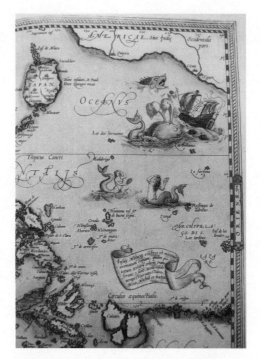

34. Detail from map of Asia in Abraham Ortelius's *Theatrum Orbis Terrarum*, Antwerp, 1570. Courtesy of the Library of Congress, Washington, D.C.

Belgium, and Germany. In 1589, twenty-two maps were produced including Italy, Slovakia, and Greece. Mercator was working on a collection of maps, which he called an atlas, when he died in 1594, but his son Rumold and grandsons Michael, Gerardus, and Johann completed the atlas and published it in 1595. The atlas is in part a memorial to Mercator; the introduction includes a verse to him and cites his biographical details in glowing terms. Its full title is *Atlas sive cosmographicae meditationes de fabrica mundi et fabricati figura* (Atlas, or cosmographical meditations upon the creation of the universe, and the universe as created). It is best known as Mercator's *Atlas*.

Mercator was born on March 5, 1512, in the small town of Rupelmonde near Antwerp. He came from a modest household and his given name was Gerhard Kremer. He later Latinized his name to Mercator, a common practice among European scholars because it referenced the classical learning of the ancient Romans. He studied at Louvain in 1530. Like so many other cosmographers of the time, he was deeply religious; he sustained an interest in theology throughout his life. He wrote theological texts including commentaries on the letters of Saint Paul, on the prophet Ezekiel, and on various biblical chronologies. Through his theology and his more overtly scientific work, Mercator sought to discover the glory of God in the "fabrick of the universe."

Mercator was caught up in the religious controversies of the day. In 1544 he was accused of heresy, arrested, and imprisoned. Among those who were charged along with him, two were burned at the stake, one was beheaded, and one was buried alive. Mercator was luckier. Supported by friends and university connections, he was released. He later moved to the more tolerant setting of Duisberg, where he taught mathematics at the gymnasium.

Mercator was well-connected to other European cosmographers. He knew John Dee and Abraham Ortelius. He learned geography, astronomy, cartography, and surveying from Gemma Frisius. He had a wide range of practical skills including mapmaking, italic writing, engraving, and globe making. In a 1540 treatise on handwriting, Mercator showed scribes how to cut the quill and the best way to hold it. He made mathematical instruments for Charles V. But he is now best known for his maps. His first map was a 1537 copper engraving map of the Holy Land. The next year he published his first world map, a double cordiform similar to Fine's 1531 double cordiform. At the request of merchants, in 1540 he produced a map of Flanders based on earlier surveys. He made surveys for royal patrons, and in 1564 he was named cosmographer to the Duke of Julich, Cleve, and Berg.

Mercator conceived an ambitious plan to write a massive cosmographical work on the creation of the world. The first piece of the work was a world chronology published in 1569. In the same year he also produced a world map on a new projection. This was the famous Mercator's projection, which represented the world as a square with the polar areas flattened out to the same extent as the equator. The projection was used by mariners because, while it magnifies the surface area of the poles, it maintains a constant compass directions. It quickly became a standard projection and was popularized in the English world by Richard Hakluyt, who published a Mercator projection world map in his *Principal Navigations* (1598–1600). The projection has had a lasting impact on the way we see the world and is still used today. As with so many other mapmakers of the time, Mercator not only experimented with new projections but also made a deferential nod to Ptolemy; he published Ptolemaic maps in 1578.

Mercator's 1595 *Atlas* was foreshadowed by his *Atlas of Europe*, produced around 1570–72. The Crown Prince of Cleves asked Mercator to prepare maps for his tour of Europe. Mercator used his existing wall maps to produce a book of maps. The initial plan was to produce an atlas of almost one hundred maps, but only seventeen are known to exist: nine of them are from his 1554 wall map of Europe; five maps are from his 1564 wall map of the British Isles, including Ireland, England, Cornwall, Scotland, Hebrides, and Orkneys; and the map of Europe is extracted from his 1569 world map. There are also two manuscript maps of Tyrol and Lombardy produced around 1570.

The 1595 *Atlas* is a large book in page size and overall heft. Not so much a coffee table book, it is almost a small coffee table. The frontispiece is of the god Atlas with the world between his hands; one bare hand covers part of the globe, while the other hand measures it with a divider. It has all the elements of magus iconography: the world between the hands, plotted and gridded, brought down from the spheres, the fantasy of global comprehension. We are not exactly sure why Mercator called upon the mythic figure of Atlas to represent his book of maps. Perhaps it was because Atlas, a historic man turned into a god by his deeds, suggested the cosmographer as an atlas-like figure. Whatever the reason, the name

35. Frontispiece of Gerardus Mercator's *Atlas sive cosmographicae,* Duisberg, 1595; RB 238463. This item is reproduced by permission of The Huntington Library, San Marino, California.

stuck and by the end of the seventeenth century the term atlas was the most widely used term for a book of maps. It still is.

In the introduction to the maps there is a thirty-page section titled "The Booke of the Creation and Fabrick of the World," in which Mercator writes of "God the author" and argues that to know the cosmos is to know the infinite wisdom of God. For Mercator and many of his contemporaries, cosmography revealed God's wonder. After the religious tone is set, there is "An introduction to Universal Geography," in which the usual distinction is made between cosmography, geography, and chorography (region or country). Topography, the writing of place, is also included. One decorative page references the classical sources of Marinus and Ptolemy.

The first map in the atlas is a double hemispheric map of the world. The southern latitudes are sketchy: New Guinea is depicted, but Australia has yet to appear from a *polar terra australis.*

36. Decorative page in Gerardus Mercator's *Atlas sive cosmographicae*, Duisberg, 1595. Courtesy of the Library of Congress, Washington, D.C.

There is limited border decoration. There are maps of Europe, Africa, Asia (with the spice islands exaggerated in size, reflecting their economic significance to Europeans), and the North Pole. More detailed maps follow. Each is prefaced by a note on latitude and longitude and general geographical introduction, and some of the maps have index tables giving the latitude and longitude of selected cities and towns. The maps are all gridded with uniform lines of latitude and longitude. A variety of projections are employed. This is a text with a surprisingly modern "scientific" look: neutral borders, few nymphs, mainly quiet decoration. The key uses literal symbols. Relief is shown as small hills, towns are de-

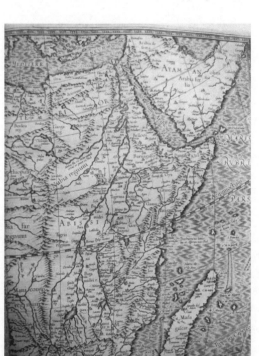

37. Detail from map of Africa in Gerardus Mercator's *Atlas sive cosmographicae*, Duisberg, 1595. Courtesy of the Library of Congress, Washington, D.C.

picted by buildings, while forests are signified by little tree icons. In some versions that I have seen, the quietly dignified plates are the staging post for some incredible flights of fancy by imaginative colorists.

Despite its claim to universality, the atlas is biased toward Europe. Most of the maps are of European regions, and there are no maps of the New World.

Around 1604 the plates were sold to Jodocus Hondius (1563–1611), who was born in Flanders but worked in London as an engraver, a mapmaker, and in the "art of cosmographie" before returning to Amsterdam. After his death, his sons-in-law, Henry and Jan Jansson, published twenty-nine editions from 1609 to 1641, as well as twenty-five editions of the cheaper pocket version, the *Atlas Minor*, from 1607 to 1638. Versions were published in Latin, French, German, Dutch, English, and Turkish. The 1630 French edition of the *Atlas Minor* had 643 pages of text. An English version, which was first published in 1636 by Henry Hondius, is even larger than the original. The title was shortened to *Atlas; or A geographicke description of the Regions, Countries and Kingdomes of the world*. The volume had brief biographies of "The life of renowned and illustrious cosmographer Gerard Mercator" and "the life of famous and excellent cosmographer Jodocus Hondius." In one illustration both Mercator and Hondius are sitting at a table together, both with dividers in their hands, plotting a globe of the world.

In the later versions, the atlas ballooned to massive proportions. The English version published in 1636 has more than two hundred maps. The mapping of the world had broadened and deepened.[4]

Atlases continue to be an important part of our view of the world. They are a standard part of every serious library, a regular school text. They may have shrunk in size from Mercator's first atlas, but there is still the same attempt to encompass the world in one text. The project begun by Ortelius and Mercator continues to this day.

The National Atlas

If the world atlas encompasses the world, then the national atlas embodies the nation-state. National mappings developed in the sixteenth century at the intersection of cartographic improvements and political necessities. As nation-states emerged from empires and as national identity was sharpened in conflict with competing countries, there was a perceived need for forms of national representation. National atlases give shape to the nation.

Through the sixteenth century, national mappings developed as geography and chorography become tied to the project of the nation-state in different parts of Europe. The mappings came in a variety of forms. In Scotland between 1583 and 1596, Timothy Pont produced very detailed manuscript maps. The National Library of Scotland currently has seventy-seven of them.[5] At around the same time, the first national atlas of France was being made, *Le Theatre Francois*, which was published in Tours in 1594. After a popular revolt, King Henry left Paris in 1589 for Tours, where a court and central administration were established. A bookseller, Maurice Bouguereau, hired Flemish engraver Gabriel Tavernier to transfer onto copperplates maps of France that he obtained from different sources.[6] These maps were collated together and dedicated to the king. The country was covered in forty-eight maps, many of them with inserts of town plans. The various uses of the atlas were cited as a "Theatre of the various Provinces of Your realm, for the pleasure of learned men, for the use of martial men and of the King's tax gatherers and treasurers and for the guidance of merchants." In the first edition of this work, the picture of Henry covered the map of France. Flick back the portrait and underneath is the map of France: king and country, nation and sovereign, the body of the country and the head of the king. The iconography is clear and direct.

The first national atlas in Europe was produced by Christopher Saxton in 1579. Saxton drew upon the earlier work of Cuningham, whose 1559 *Cosmographical Glasse* showed how to make a map of England and who printed the first engraved town plan. Saxton's work represented not just a technical accomplishment. The mapping was uniquely connected to political ends. It is important to understand the wider context. Since the time of Henry VIII's break with Rome, England had been rent by religious factionalism. The Crown had shifted from the Protestant evangelism of Edward VI's brief reign (1547–53) to the militant Catholicism of Mary Tudor (1553–58) and back to the Protestantism of Elizabeth's reign (1558–1603), which was beset by tensions between Protestants and

38. The city of Limoges from Maurice Bouguereau's *Le Theater Francois*, Tours, 1594. Courtesy of the Library of Congress, Washington, D.C.

Catholics at home, international rivalry abroad, and the growing enmity of Spain. The disputes of the Reformation and Counter Reformation and overt Spanish hostility marked the Elizabethan era and the makings of Saxton's atlas.[7]

There had been mappings of England before Saxton. The Benedictine scribe Matthew Paris drew the first map of England in the thirteenth century, based on much earlier, possibly Roman, itinerary maps; in addition, the "Gough" map of Britain (ca. 1360), with its elaborate road system and urban hierarchy, was one of the best maps of medieval Europe. There had also been ambitious plans to map the whole country. Nicolas Kratzer, astronomer to Henry VIII, wrote to Albrecht Durer about his plans to make a map of England based on extensive fieldwork. In 1561, John Rudd (ca. 1498–1579), at the request of Elizabeth, was granted two years' pay to prepare a map of the country, but the product never materialized.

Christopher Saxton was born around 1542–44 in Yorkshire. Little is known about the details of his life, but we do know that he became a surveyor. The dissolution of the monasteries had created a large pool of commodified land, as the vast estates of the religious orders became part of the commercial land market. The enclosure of land, a process of privatization of public lands into private hands—especially the hands of the already wealthy—was also creating a more capitalistic land market that needed to be mapped and surveyed. Saxton was employed as an estate surveyor by private landowners and by official courts investigating disputed lands. Saxton became well connected and prospered in the shifting quagmire that was Elizabethan England. In 1573 he was chosen by Thomas Seckford to survey and map the counties of England and Wales. Seckford was a state functionary; he was a master of the Court of Requests and later, surveyor of the Court of Wards. His boss was William Cecil, later Lord Burghley, who was one of the most powerful people in the land—he was the queen's secretary of state, a privy councilor, and later lord treasurer.

Burghley took an active interest in Saxton's work. He was concerned with surveying the realm for many reasons, the most important one being the need for accurate and regular surveillance. Burghley had an extensive cartographic collection, promoted various mappings including a survey of southern Ireland by Robert Lythe, and wrote detailed annotations on many of his maps. On one of his maps he entered the names of Catholic recusants beside their homes. With the Saxton survey, he saw each plate as soon as it was engraved and bound them into his own atlas, to which he added notes and other maps. This atlas received lots of annotations before and during the attempted invasion of England by the Spanish Armada in 1588. On Saxton's map of Northumberland, now in the British Library, Burghley annotated, in his spiky handwriting, the number of horses that the local gentry could raise in the event of war. Burghley was always on the lookout for enemies at home and abroad. He knew that sound geographical intelligence, like the Saxton survey, was always needed.

Saxton was officially appointed by the queen to undertake the survey of the country in July 1573. Seckford paid all of Saxton's costs, but it really was a state project promoted by Burghley. In return, Saxton received official favor. In 1573 he was given an estate in Suffolk, formerly monastery lands. In 1574 he got another grant of land in London and in 1579 he was given a formal coat of arms. The loyal functionary had achieved the pinnacle of establishment acceptance. When his arms were granted, the official proclamation noted, "by special direccion & commendment from the Queenes Majesty hath endeavoured to make a perfect Geographicall discripcion of all the seurall Shires and Counties within this Realme." His crest was a hand and arm holding a pair of partly open compasses.

Saxton's survey lasted from 1574 to 1578. He began in Norfolk and the south Midlands; then he moved to Essex, east Midlands, and then into the North. All the English counties were done by 1577 and all the Welsh counties by 1578. Saxton's survey was an official mission. He was given an open letter to local officials "to see him conducted unto any towre Castle highe place or hill to view the countrey." The aim was not so much to identify latitude and longitude, which do not appear on his atlas maps; it was to identify the lay of the land. He was taken up to high points in various counties, accompanied by two or three "honest men."

Saxton's maps were surveillances of the national territory. There was a need for surveillance. Catholics were plotting against the queen. In 1570 the pope, in a declaration that can only be described as a form of Catholic *fatwa*, had given a free pass to heaven to any Catholic able to kill Elizabeth. The Spanish posed a continual threat. In 1567 the Spanish army in the Netherlands was dangerously close. In 1569 the Privy Council called on counties to have a general muster of all able-bodied males older than sixteen years and to create a beacon system of bonfires on hills to alert of foreign invasion. The beacons sites were probably the high points that Saxton was taken to by the "honest men." The route of the survey reflected strategic considerations. Saxton surveyed the vulnerable south coast counties before the less strategic northern counties and Wales.

The very use of counties as the principal unit reflected the chain of command. Counties were a military unit of allegiance to the Crown; the power flowed from the Crown to the lord lieutenants of the county down to local justices of the peace. The counties were responsible for military musters. Counties were not the rather quaint divisions they are today. Rather, they were the principal units in the spatial, military, and political administration of the realm. Saxton used this basic military-political unit as his template.

Having surveyed all the counties of England and Wales, Saxton brought them together in an atlas. The 1579 *Atlas of England and Wales* has a frontispiece of Elizabeth, showing her as the patron of geography and astronomy. The first map is a map of England with no grid, which is followed by double-paged maps of the individual counties. Each map exhibits the royal coat of arms, Seckford's motto and coat of arms, and Christopher Saxton's name—often

39. Map of Dorset in Christopher Saxton's *Atlas of England and Wales*, London, 1579. Courtesy of Newbery Library, Chicago.

90 ❧ *Making Space*

beside the scale diagram of a set of dividers. The maps show rivers, towns, enclosed forests, and the houses and castles of local gentry. Relief is shown in diagrammatic form as little molehills, roughly to scale; much larger molehills appear in more mountainous counties in the North and Wales. No roads are shown, suggesting that the survey was done quickly and rather crudely. Roads were probably invisible from the high points and their ac-

40. Detail from map of Denbeigh in Christopher Saxton's *Atlas of England and Wales*, London, 1579.
Courtesy of the Library of Congress, Washington, D.C.

curate mapping would have taken up too much valuable time. Roads were not shown accurately on English maps until the work of John Ogilby in 1675.

The county maps in the 1579 *Atlas* were a vital geographical intelligence. There was no text to accompany the maps. However, the stamping of the royal coat of arms on each map speaks volumes about the connection between the body of the country and the figure of the royal personage. The coats of arms of England's ancient nobility were added to later versions—beginning in 1590—to give the book a more national character.

Very much a product of its time, the atlas was to become a landmark in English cartography. The atlas was published throughout the seventeenth century and was the basis for most county maps until 1650. The final edition was published in 1730. The county maps were reproduced in later editions of Camden's *Britannia* (1607) as well as Speed's *Theatre of the British Empire*. Camden called Saxton the "optimus Chorographus." Saxton provided the basis for English maps for almost one hundred years and provided the most complete survey of the country until the creation of the Ordnance Survey in 1791.

Saxton also used the work of his survey to create a large map of England and Wales in 1583, again under Seckford's patronage. The map was printed on twenty sheets at the generous scale of seven miles to an inch on a trapezoid projection. It was the earliest wall map that was engraved and printed in England. The map was regularly sold until 1795 and was used in various forms by such later Low Country mapmakers as Plancius and Hondius. It was also used as decoration. A visitor to Burghley's house in 1592 noted a large wall map showing the kingdom; it was probably a copy of Saxton's map.

Saxton's survey and the resulting atlas and large wall map have been put in a wider context by Richard Helgerson in his 1992 book, *The Elizabethan Writing of England*. Helgerson argues that beginning in the 1560s there was a self-conscious writing of England that included Spenser's *Faerie Queen*, Camden's *Britannia*, Speed's

Theatre, Shakespeare's historical plays, Haklyut's *Principal Navigations*, Coke's *Institutes of the Laws of England*, Drayton's *Poly-Olbion*, and Hooker's *Laws of Ecclesiastical Polity*. Helgerson calls it a concentrated generational subject, as different discursive communities were linked in a similar project of national representation.[8]

There was a tension in the project between absolutists and more democratic notions of the country. Some of these tensions were revealed in Saxton's work. Representing the land meant different things to the different groups. The absolutist claim was embodied in the royal coat of arms shown on every county map, but the maps were also places where the country gentry could find their manors, monuments, and pedigrees. Therefore, there was a tension between chorography and royal absolutism. Royal possession was partly undermined by showing the particularities of the local gentry.

The creation of a national spatial consciousness can also be seen in two of the other works mentioned by Helgerson. William Camden's *Britannia* was first published in 1586. It is subtitled a *Chorographical Description of the most flourishing kingdomes, England Scotland Ireland*. Part local history and part chorography, the book became very influential. Camden noted that a common complaint was that he did not include maps, which are important because they "doe allure the eies of pleasant portraiture, and are the best directions in Geographicall studies." Later editions, from 1607 onward, use the maps of Saxton and John Norden, described as "most skilfull Chorographers." A 1637 edition proudly proclaimed it is "Beautified with Mappes of The Shires of England."

The poet Michael Drayton (1563–1631) wrote his poem *Poly-Olbion* as a poetic chorography. First published in 1613, its famous title page has a picture of a female allegory of the country draped in a map cloak. It is both a monarchical image and a land-centered view of the country. The nation is less the royal personage and more the land itself. *Poly-Olbion* is a fifteen-thousand-line national poem that begins in Devon and Cornwall with the line "Of albion's glorious Isle, the wonders whilst I write" and thirty songs

Mapping the World ✣ 93

41. Frontispiece of Michael Drayton's *Poly-Olbion*, London, 1613; RB 494928. This item is reproduced by permission of The Huntington Library, San Marino, California.

later ends in the far north of Westmoreland and Cumberland. Each part of the country has a song and map full of nymphs and choirs. There is considerable debate about how to read the poem.[9] What is clear, however, is the importance of place and cartographic representation. *Poly-Olbion* is a poetic chorography that conjures up images of the country.

The later Elizabethan and Stuart periods in England were awash in cartographic representation of the country and the nation. Not all the projects were as successful as Saxton's. John Norden, for example, provides an example of a national mapping project that never quite materialized.

John Norden was born in Somerset in 1548, probably the son of a yeoman. He made a living from surveying the lands of the former monasteries. He developed the idea for a national geography around 1586–89. At that time he was surveying former monastic lands given to Sir Thomas Heneage in Northants. By 1591 he had completed a map of Northamptonshire and sent it off to Burghley accompanied by an introduction from Heneage and a letter asking to "consider whether it might be expedient that the most principall townes, Cyties and castles within eurye Shire, should be breefly and expertly plotted out." Burghley must have liked and supported the proposal. It seems that he could never get enough surveys of the country because the next year Norden successfully petitioned the Privy Council, which granted him a ten-year monopoly for printing his survey work.

Like Saxton before him, Norden was given a royal pass. A Privy Council order of 1593 noted, "John Norden gent is authorised and appointed by her Maiesty to trauayl into the several counties of the realme of England and Wales and to make perfect descripciones, chartes and mappes."

Norden planned to write a comprehensive geography of the country. Its title was *Speculum Britanniae* (Mirror of Britain). He attempted to join Camden's historiography and Saxton's chorography, combining topography and historical interest in maps useful for administrators, merchants, and the educated classes. The project was never completed and information on only two counties was published during his lifetime. It started well enough. *An historicall and chorographicall description of Middlesex* was published in 1593. It is a small book dedicated to Elizabeth and Burghley containing a map of Middlesex plus two maps of London. The Middlesex map uses an index system of letters (a, b, c, etc.) and numbers (1, 2, 3, etc.), much as our modern maps use, to identify points of interest. The maps show roads, towns, hamlets, and enclosures. The accompanying text provides a long history from earliest times, discussion of a variety of topics including soil and fertility, and lists of market towns. It is a geographical and historical inventory, naming and recording, fixing places in time as well as space. Norden was giving order to the national geography. In 1595 he produced a manuscript atlas of the four counties of Middlesex, Essex, Surrey, and Hampshire.

To sustain the project Norden needed money and patronage. He ran out of both. Burghley died in 1598, and in that year Norden published *An historicall and chorographicall description of Hertfordshire* at his own expense. In this book he wrote an open letter to the queen for help, "Onlie your Maiesties princelie fauor is my hope, without which O my selfe must miserablie perish, my familie in penurie and the work unperformed, whiche being effected, shall be profitble and a glorie to this your most admired Empire." The flattery did not work, but Norden, never one to give up, approached the new king in 1604, asking to do a survey of the king's lands and a "redescription of Shires of Englande." The "redescrip-

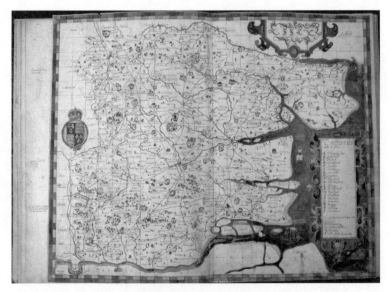

42. John Norden's manuscript map of Essex, 1595; K47541 ff. 9–10. By permission of the British Library, London.

tion" never happened, but Norden achieved some royal favor. In January 1605 he received a lifetime appointment as surveyor to the Duchy of Cornwall. I will say more of his contributions to surveying in chapter 6.

While Norden's grand project never came to pass, he left an important legacy. His books on Hertfordshire and Middlesex raised the level of technical representation. He was aware of the variation in the mile used in different parts of the country, from 1120 yards in some places to 1760 yards in others. He used the 1760-yard mile. He also raised the issue of the singular naming of places at a time when names varied—for example, variant spellings of Bury or Berye, Ley or Leigh. Writing at a time when the names of the same place and even the length of miles varied around the country, Norden worked to establish a national metric and a formalized system of naming in his attempt to create a truly national geography. In 1625 he also published *An Intended Guyde for English Travailers*, in which he provided a table of distances between towns and cities in each county. The book represents a national space bound by a network of connections.

Where Norden failed in his ambitious attempt to write a national geography, John Speed was more successful. John Speed (1551/52–1629) was born in Cheshire and worked in London as a merchant tailor until 1598, when he was given a sinecure at the custom house by Sir Fulke Greville, enabling him to devote his time to historical and geographical studies. In later life he also wrote religious pieces. Speed was interested in both the history and the geography of Britain. There was intense interest in both subjects in the Elizabethan and early Stuart era. The search for historical antecedents and the creation of national geography were two elements of a self-conscious national representation that marked this period. In his *The Theatre of the Empire of Great Britaine*, first published 1611–12, Speed created a lavish national atlas.

Speed consciously adopted Ortelius's term *theatre*. Ortelius had visited England and shared his interests in antiquities and geography with both Speed and Camden. Speed's *Theatre* was a large book dedicated in effusive terms to King James, "restorer of the

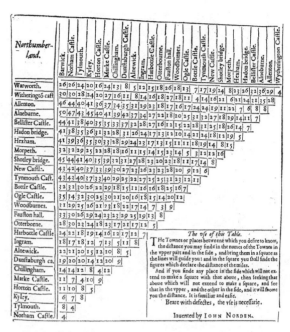

43. Table of distances in John Norden's *An Intended Guyde for English Travailers*, London, 1625. Courtesy of the Library of Congress, Washington, D.C.

British name." A self-conscious patriotism is evident; Speed refers to the country as "the very Eden of Europe," offers his work "upon the altar of love to my country," and notes that he has a "zeal for my country's glory."

The *Theatre* contains general maps of Britain, Scotland, England, and Ireland as well as detailed county maps of England. The county maps draw extensively on the work of Norden and Saxton; they show woods, towns, coats of arms of local gentry, and sometimes historical sites. The county map of Wiltshire, for example, contains an illustration of Stonehenge. Accompanying the map are brief descriptions of location, history, and geography. While Speed drew heavily upon others, he also mapped fifty of the seventy-three town plan maps shown in the *Theatre*. He calculated many of the measurements himself by pacing out the city streets.

44. Map of Cambridgeshire in John Speed's *The Theatre of the Empire of Great Britaine,* London, 1611–12. Courtesy of the Library of Congress, Washington, D.C.

The city plans are accurate depictions but they are also national advertisements that depict sites of peace and industry to glorify the country. In a nod of deference to the new Scottish king of Britain, James VI, the map of Britain contains town maps of both London and Edinburgh.

The *Theatre* accompanied Speed's 1611 *History of Great Britaine,* which is basically a long historiography of the kings of England from Norman times leading up to Elizabeth and James. For the more recent monarchies it offers a blow-by-blow account, surprisingly modern in its narrative drive. Copies of the *History* and *Theatre* were sometimes bound together in one volume.

The *Theatre* contains maps with elaborate borders that depict different social strata. The map of England, for example, has pictures of a "nobleman, gentleman, citizen and countryman." On the other side are depicted a "lady, gentlewomen, citizen's wife and countrywoman." Thus the *Theatre* depicts social location as well as spatial location. The map of Scotland includes as border decoration "Scotsman and Scotswoman" as well as "Highlandman and Highlandwoman." These are important social divisions because the former are part of the empire and the latter are on the margins. This geopolitical distinction is reinforced in the border decorations of the map of Ireland, with its telling division of "gentleman, civil Irish and wild Irish."

The *Theatre* was published in a variety of editions. A Latin edition was published in 1616 for a wider European market. The empire that Speed alluded to in the original title was England, Ireland, Scotland, and Wales. Subsequent editions of Speed's *Theatre* included parts of what we now call the British Empire. The English 1676 version, for example, contained maps of "His Majesty's Dominions" in New England, New York, Carolina, Florida, Virginia, Maryland, Jamaica, and Barbados. As the *Theatre* expanded to include the newer colonial holdings, the national atlas became an imperial atlas.

Two things happened in England—and indeed in many European countries—in the course of the sixteenth century. The first was a cartographic revolution. In 1500 maps were little known and little understood; by 1600 they were familiar objects. The first printed map appeared in England in only 1535; it showed the biblical scene of Exodus. By the end of the century, maps were part of national life as a widespread cartographic literacy was created. Saxton's county maps, for example, appeared on the backs of playing cards and, in a famous portrait of Elizabeth (ca. 1592) by Marcus Gheeraerts the Younger, the queen is depicted standing on Saxton's map of England and Wales—map of the country, body of the monarch in one compelling image.

The second major change during the century was the creation of a more overt national consciousness. Nationalism is a social creation, which develops from a combination of schooling, religion, rituals, stories and holidays, history and geography, among other things. One essential element was, and continues to be, a recognizable image of the country. The two processes of cartographical representation and emerging national identity are closely connected.

45. Marcus Gheeraerts, *Portrait of Queen Elizabeth*, ca. 1592, oil on canvas. Courtesy of the National Portrait Gallery, London.

The maps of the sixteenth century that are integrated in national atlases are not merely depictions of places; they are also generators of national consciousness. But the development of a national consciousness is never a simple process. The definition of the nation-state is always a point of dispute rather than an established fact. The creation of a national atlas, like the construction of national identity, is always problematic, never fixed, a source of continual debate and struggle. The national atlases can be and are read in very different ways—signs of absolutist power as well as indicators of a wider, broader national community. Mapping the country is not a technical exercise, devoid of political meaning. When the country is mapped and individual counties are brought together in one volume, national identity is given cartographic form, a physical and textual presence that people can imbibe and read. The national atlas is an administrative/surveillance device that also embodies the nation-state and reinforces national consciousness. A project designed to survey the nation also helps to make the nation.

Mapping the City

Cities were the backbone of economic growth and social change in early modern Europe. The creation of a money economy, the emergence of a merchant class, and the growth of manufacturing and trade were all intimately linked to the city. The European merchant city was an important entity in its own right, often with distinct city charters and forms of government. City-states were important and distinct economic and political units conferring special economic privileges and social freedoms to their citizens. The city was an important source of identity.

Cities were, and still are, shown from three main views: as a prospect (the view from the side), a plan (the view from directly above), and a bird's-eye view (the view from high above but to the side). In all these urban representations, there is no hard and fast divide between cartography and art, mapmaker and artist.

Manuscript copies of Ptolemy's *Geography* produced in ca. 1456, 1469, and 1472 all contained city maps. The 1472 manuscript, made for the Duke of Urbino, contained maps by artist Pietro del Massaio of Milan, Venice, Florence, Rome, Constantinople, Damascus, Jerusalem, Alexandria, and Cairo. They are all bird's-eye views done in watercolor and ink on vellum.

Formal mappings of the city coincided with the development of perspective. Alberti drew upon the grid of Ptolemy to formally express the principles of perspective. In his writings on painting (1436), architecture (1452), and sculpture (1464), Alberti provided discourses on the representation of space. His *Descriptio Urbis Romae,* probably done in the 1440s, provided the coordinates for a scaled map of Rome. The buildings are drawn in accurate relation to each other in space on a circular plan. Each building is plotted in a coordinate system. Leonardo da Vinci mapped the city of Imola on the circular model used by Alberti. He then made working sketches based on his own pacing out of measurements in the city. His 1502 map of Imola, an exemplar case of both beauty and accuracy, provides a detailed plan of the city. Leonardo's plan map was the exception. Most city "maps" in the sixteenth century were prospect or bird's-eye views rather than plans. The oblique view allowed a rendition of the buildings and walls. The vertical plan, in contrast, flattened out the built form. Cities were not so much mapped as represented, and the location and relative size of buildings was an important consideration. Such representations were often used to memorialize and celebrate a city. Buildings hidden by the vertical plan were highlighted by the prospect and bird's-eye view.

Printed images of cities appeared soon after the invention of printing. A popular early book was *Peregrinatio in terram sanctam*,

46. Leonardo da Vinci's *Map of Imola*, ca. 1502, Royal Library, Windsor. The Royal Collection, © 2003, Her Majesty Queen Elizabeth II.

an early "travels in the Holy Land" book published in Mainz in 1486. It documented a pilgrimage in 1483–84 to Jerusalem led by the Dean of Mainz Cathedral. One of the party was Erhard Reuwich, who drew illustrations, including cityscapes of the places visited en route. It is one of the first books with illustrations drawn from life. The book has detailed, relatively accurate city views. In the *Nuremberg Chronicle,* for example, there are illustrations of close to one hundred cities.

One of the largest and earliest printed bird's-eye view maps is Jacopo de' Barbari's 1500 illustration of Venice. It is a large work, measuring 135 centimeters by 282 centimeters, composed of six large sheets. The city is depicted in a wealth of detail as if one is looking down at it from the southwest. The map is as much a work

47. View of Lyon in Hartmann Schedel's *Liber cronicarum,* Nuremberg, 1493; leaf 78r, woodcut. This item is reproduced by permission of The Huntington Library, San Marino, California.

of art as a cartographic monument. The author was a painter and printmaker, born in Venice around 1450. He worked in both Venice and Nuremberg and served as a court artist in various small European courts. The project was devised by a Nuremberg businessman and Venice resident, Anton Kolb, who in 1500 petitioned the Venetian government for monopoly publishing rights. The view of Venice is a studio fabrication assembled from many drawings made in different parts of the city over the course of three years.

Many other cities were soon after represented. Francesco Rosselli produced prospects of Pisa, Rome, Constantinople, and Florence in the 1490s; an anonymous woodcut prospect of Antwerp appeared in 1515; a bird's-eye view of Augsburg was produced 1521 by Jorg Seld; and a woodcut prospect of Amsterdam appeared in 1544 by Cornelis Antoniszoon. Hans Lautensack's view of Nuremberg, produced in 1552, showed a wealth of accurate detail and included inserts that stressed the city's antiquity and appealed to God's favor for municipal success. City maps celebrated the city. They also had a talismanic quality; representations of the city were used to invoke good fortune and prosperity. The detailed urban maps were a source of civic pride and were often used to illustrate local chronicles and proclaim the identity and prestige of a city. The level of detail in some of the illustrations also suggests the urban map as panopticon, a form of cartographic surveillance.

Many of the early city prospects were used in a generic way to represent the idea of a city rather than an actual city. They look like real places with buildings and walls, but much of the detail is imaginary and the same woodcuts were used to illustrate very different cities. It was the idea of a city rather than the actuality of a particular city that seemed most important. City images served didactic rather than reporting functions. The same illustration was used to represent different cities even in many sixteenth-century texts. Munster uses town prospects in the first edition of *Cosmographia* in 1544, but they were generic rather than particular illustrations. The publication of Johannes Stumpf's *Schweizer Chronik* in 1548, with its more accurate representation of Swiss cities, prompted Munster to revise his city views in the 1550 edition of

Cosmographia. In 1548, Pedro de Medina presented to King Philip of Spain a description of every town in Spain with a panoramic woodcut. But even in the second edition of 1595 the woodcut for Seville was also used to depict both Gibraltar and Aragon.

Cities were mapped for a variety of reasons. Cities were military strongholds and strategic sites. The earliest extant urban map in England was a 1545 plan view map of Portsmouth, which was done to accompany proposals for improving the town's defense after an attack by the French earlier that year. Cities were also mapped as part of a national geographical information system, a royal seeing of the territory. Philip of Spain paid for two large urban surveys. In 1559 he ordered cartographer Jacob van Deventer to measure and draw all the towns in the Spanish Netherlands. By 1572 almost three hundred bird's-eye views had been completed. In the 1560s Philip gave a similar commission for the towns of Spain to the Dutch artist Anton van den Wyngaerde, who produced more than sixty prospect pictures.

By the last third of the sixteenth century, there was a considerable stock of urban maps and images. They had been drawn for a variety of reasons: civic pride; celebrations of specific events, such as the colossal prospect of Cologne by Anton Woensam drawn in 1531 on the election of Ferdinand of Austria as king of the Romans; military surveillance; and part of national inventories. Compilations of city maps and prospects were published in 1551 and 1567, but the first city atlas was the *Civitates Orbis Terrarum* by Georg Braun and Frans Hogenberg. One volume was published in 1572 but it became so popular that by 1617 the work consisted of six volumes with more than 363 urban views. Forty-six editions were produced in Latin, German, and French. The success of the atlas started a fashion that was to last into the eighteenth century. Matthaus Merian published *Theatrum Europaeum* in 1640 in a twenty-one-volume work consisting of 500 plates, most of them from the *Civitates,* and 173 newly engraved plates. Jan Jansson purchased 363 of the *Civitates* plates in 1653 as the basis for an eight-volume work first published in 1657. After the *Civitates,* even national atlases depicted cityscapes. John Speed's *Theatre,* pub-

48. Map of Groningen in Georg Braun's *Civitates Orbis Terrarum*, vol. 2, Cologne, 1612; RB 180544. This item is reproduced by permission of The Huntington Library, San Marino, California.

lished in 1611–12, contained more than seventy views of British cities, visited, as Speed noted, "by mine owne travels."

The first volume of the *Civitates* was published in Cologne, edited by Georg Braun, and engraved by Frans Hogenberg; it contained prospects, bird's-eye views, and plans of cities from all over the world. Braun (1541–1622) was a Cologne cleric who obtained maps and drawings from cartographers throughout Europe, commissioned work, and wrote most of the descriptions. Hogenberg (1535–90) was an engraver who had worked on Ortelius's *Theatrum*. The *Civitates* was intended as a companion volume to the *Theatrum*. In a letter to Ortelius, Braun wrote that the atlas was meant to appeal to the lettered and unlettered. The Latin script appealed to the educated but the views also made it accessible to the unversed. Sometimes two sets of names were used for one city— the Latin and the vernacular—so that the literate, but less scholarly, could see the name of their city.

Civitates drew from a variety of sources, including Sebastian Munster, Johannes Stumpf, Jacob van Deventer, and artist and illuminator Georg Hoefnagel. Braun appealed to his readers to send him views and so the atlas grew with successive editions. The map of London, for example, shown in Braun and Hogenberg was based on the first surviving printed map of the city, done in 1553–59. It is referred to as the Copperplate Map. Completed at a scale of 25–34 inches to one mile, it measures a grand 3.8 by 7.5 feet and covers fifteen separate plates. We are not sure exactly why the map was drawn, but perhaps it was made for a royal event such as a coronation. This map was the basis for the Braun and Hogenberg map of 1572, engraved by Hogenberg at a reduced scale of 6.5 inches to one mile.

Civitates provides us with a comprehensive collection of sixteenth-century urban views. Many of the views depict people in local dress. In the introduction Braun noted that representing the human figures would keep the Turks from viewing the maps because they were forbidden by their faith, he believed, from looking at human figures. The remark reflects a fear of the Turks gaining knowledge of vulnerable European cities.

Braun stipulated that the towns should be drawn so that the viewer could look into all the roads and see all the buildings. In some cities individual buildings are named. Each city has a brief written note of its history, situation, and commerce. The prospect and the bird's-eye view predominate, and even when the city is shown as a plan, buildings are shown in vertical relief.

In *Civitates* the city is both displayed and bounded. In almost all of the images, the city walls figure largely. Cities were often fiercely independent, being home to independent power centers, princes and prelates, guilds and town councils. Looking through the atlas at the many pictures of cities, one gains a very strong sense of cities standing apart as separate communities, reinforced by walls and battlements.

The images also reflect the grandeur, wealth, and power of the city. The city is not just represented but is also celebrated. Many of the urban maps and views were made to evoke and represent civic pride. The maps often adorned civic offices. In the atlas some of the cities include the loving detail of individual street names, buildings, and churches. In many of the images, the city comes alive, animated by height and dimensionality; it is clearly a complex form of representation meant to honor the city. The atlas rejoices in the urban condition.

Collectively, the images provide a comprehensive view of urban life in the Renaissance. They also indicate a world economy tied together through trade and linkages between urban centers. Aden, Peking, Cuzco, Goa, Mombassa, and Tangiers as well as other cities around the world are represented. The global reach of mercantile capitalism and European colonization is evident in the range of cities represented in the *Civitates*. While the cities are depicted separately, the effect of the compilation is to reveal a world economy of urban nodes and a trading world of connected cities.

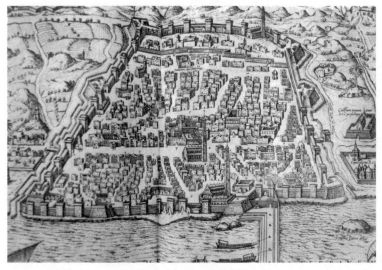

49. Detail from map of Algiers in Georg Braun's *Civitates Orbis Terrarum*, vol. 2, Cologne, 1612. Courtesy of the Library of Congress, Washington, D.C.

5

NAVIGATING THE SEAS

In 1542, Jean Rotz, a native of Dieppe, France, in the employ of Henry VIII of England, wrote,

> the world was already well supplied with the common sort of maritime charts, I formed the opinion that it would be best to make a book for him [Henry VIII] containing all hydrography or marine science; both because I had never seen such a thing, and also because it would be more useful and profitable, more significant and easier, and more convenient to handle than a single chart four or five yards in length.[1]

Rotz's *Boke of Idrography*, presented to Henry in 1542, was a sumptuous work befitting a king, all gilt and vellum and beautifully illustrated. The book also suited a Renaissance monarch because it embodied the latest learning and discoveries; it included representations of the recently discovered New World and Far East Asia. The book is full of maritime charts of coastal areas around the world.

The making of space in the sixteenth century involved the charting of the oceans as well as the mapping of the land. In the late fifteenth and early sixteenth centuries, European merchants broadened and deepened trade routes throughout the world while the imperial projects of states also began to cast a wider net around the globe. This longer-distance movement depended on accurate spatial knowledge. The mapping of the seas and oceans was an applied cosmography, deeply embedded in political and economic interests. The improved charting of the oceans was intimately connected to the extension of trade routes and imperial ventures.

Rotz's book was a portolan atlas. The term *portolan* comes from the Italian word *portolano*, a pilot's book that contained written information on sailing directions, tides, and anchorages. The terms *rutter, roteiros,* and *routiers* was also used by, respectively, the English, Portuguese, and French. An Italian portolan of the early fifteenth century contains these directions: "From Carminar to Cartegena is 20 miles northeast by east. Cartagena is a good port at all seasons, before which port there are islands a mile distant. You may pass between any of these islands and the mainland which forms a point. As you enter the port, beware of shoals. Sail close to the middle of the channel, but towards the northeastern shore, where you may anchor."[2] Directions such as these allowed sailors to set a course and find harbors. Portolans had been around since classical times, but the oldest-known printed portolan was published in Venice in 1490. An early, widely used text was *Le Routier de la mer,* published in Rouen during 1502–10. Written by Pierre Garcie, it covered the coasts of England, Wales, France, Spain, and Portugal. It was published in more than twenty editions and numerous translations. The first English translation was published by Robert Copland in 1528.

Portolan charts are the spatialization of the pilot book. They first appear in Europe around 1300, about the same time that the compass was initially used. The compass allowed a directional metric, and portolan charts are covered in compass lines that allow mariners to plot a course. Identifying north by compass made direction finding much more accurate. The maps are oriented toward the north because it is easier to align the compass reading that way. And so began the convention of maps with north located at the top. It is now such an expected convention that we find it difficult to imagine any other orientation.

Portolan charts are the mariner's mapping of the world. The emphasis is on the coastline, so the interiors are hazy and often left blank. The focus is the littoral between sea and land. Centers of chart making developed in maritime centers in Italy, Catalonia, and France, and chart production multiplied rapidly during the three hundred years from 1300 to 1600.

Charts were valuable items. They gave detailed spatial knowledge of how to move around the world. They contained important secrets; the first thing that Sir Francis Drake did when capturing Spanish and Portuguese ships was to take their portolan charts. He even took the chart makers. Drake took the experienced pilot Nuno da Silva as prisoner for fifteen months on his circumnavigation around the world in 1577. In 1681 the buccaneer Bartolomew Sharp captured the Spanish ship *Santa Rosario* off Guayaquil. Sharp noted in his journal, "In this prize I took a Spanish manuscript of prodigous value. It describes all the ports, harbours, bayes, sands, rock & riseing of the land & instructions how to work a ship into any port or harbour between the Latt. of 17 15 N and 57 S Latt. They were going to throw it over board but by good luck I saved it. The Spanish cried when I gott the book."[3]

Working charts received rough treatment. Carried on swaying ships, subject to saltwater and constant use, few remain. Those that have survived the centuries tend to be less the working documents for practical seamen and more the presentation copies for officials and nobles. There are a number of charts, for example, that first record the New World: the Cosa chart of 1500, the Cantino chart of 1501–2, and the King-Hamy portolan chart made in Italy around 1502. The King-Hamy chart is a parchment chart with both a detailed rendition of the Brazilian coast and the outlines of Newfoundland. South America and North America stick out as unconnected coasts, still to be connected to a more solid geographical discourse.

Portolan Atlases

Mariners made collections of sea charts for their favored routes and regular trips. Collections of portolan charts developed into portolan atlases, which developed during the sixteenth century both as everyday tools for mariners and as items of luxury consumption.

Guillaume Brouscon of Brittany made a portolan atlas in 1543. It was a small pocket book made for Breton seamen, containing charts as well as tide tables, a calender of Christian feasts, a compass card, a star clock, and tables of solar declination. The volume has charts of the coast of northwest Europe including Spain, France, England, Scotland, Ireland, Denmark, Flanders, and Germany. The charts are delicate miniatures with pictures of towns in the interior; ports are linked by weaving lines that emanate from the compass roses. It is a sophisticated handbook that could be used to calculate tides, feast and fast days, the date of Easter, and new moons. It includes navigational aids such as the calculation of latitude by the Pole Star.

The rich and powerful were purchasing and collecting more than just Ptolemy's *Geography* and the atlases of Mercator and Ortelius; they were also acquiring portolan atlases. Dieppe in France was a center for the production of luxury-model portolan atlases; a particularly fine example is the Vallard Atlas of 1547. It is a large book, with hand-drawn maps on parchment, consisting of the mapping of the world from the sea looking in to the coast. Coastal points are marked in red and black consecutively. The portolan charts, fifteen in all, grid the world in lines of latitude and compass lines. The interiors contain illustrations—camels in Arabia, elephants in Africa—that have a feel of perspective rather than a medieval flatness. Taken as a group, the charts provided a complete nautical map of the world. Individually, the charts also reveal much about local geographies, both real and imagined. The chart of North America, for example, includes Jacques Cartier in a short black coat and red hat with members of his expedition sent in 1534 by the king of France, Francis I, to search for lands and discover new routes to China. There were also two charts, *La Java* and *Terra Java*, that show the coastline of northern and eastern Australia. Although produced in France, the cartographer was either Portuguese or based this work on a Portuguese text because this atlas

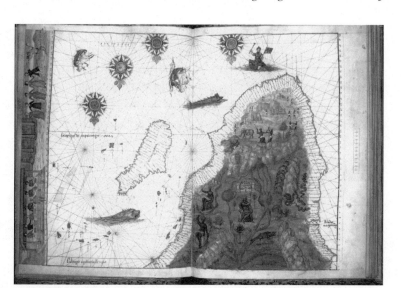

50. Portolan map of Southern Africa in Vallard Atlas, Dieppe, 1547; HM 29, map 5. This item is reproduced by permission of The Huntington Library, San Marino, California.

shows in great detail the extent of Portuguese discoveries in South America and East Asia.

The *Boke of Idrography* was a made by a member of the Dieppe school, Jean Rotz, who was a native of the French seaport (although his father was Scottish). Dieppe was an important maritime center that had also become a center of learning and scholarship hosting both Italian and Portuguese scholars. Born around 1505, Rotz may have been involved in seafaring. Around 1529 he went to Paris and learned of the work of Apian and Frisius. From 1536 to 1540 he worked on portolan charts. The king of France already had enough experts in chart making, so in 1542 Rotz went to England. Here he presented a navigational treatise to the Court and in September 1542 received his first payment as the Royal Hydrographer. Rotz was just one of the many European scholars appointed by Henry. The *Boke* is a large, Dieppe-style portolan atlas that contains instructions to navigators and tables of solar declension. The charts are oriented with south at the top of the page and the illustrations shown the opposite way up. This curious layout is best understood if we know that the atlas was meant to be laid on a table and seen from all four sides.

The Dieppe atlases were showpieces made for royals, the "Rolls Royces" of nautical mapping. A smaller and cheaper alternative, although still very expensive, were the Agnese atlases made for lesser princes and wealthy merchants. They are named after Battista Agnese, a prolific Venetian chart maker whose workshop produced more than sixty different sea atlases during the period 1536–64. These atlases are smaller than the Dieppe-school texts, but they are still beautifully executed. One example, probably made in Venice in 1544, contains a calendar, a table of declinations, a picture of an armillary sphere, a zodiac, nine sea charts, and an oval map of the world. The charts are full of vignettes of cities, forests, mountains, and rivers. The lettering along the coast is consecutively red and black to make reading easier.

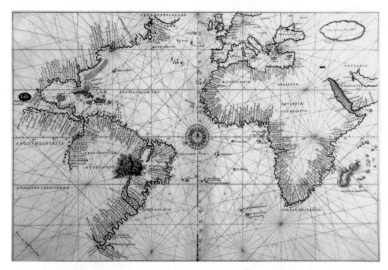

51. Map of Atlantic in *Portolan Atlas* by Battista Agnese, Italy, ca. 1544; HM 26, ff. 4v–5r. This item is reproduced by permission of The Huntington Library, San Marino, California.

Portolan atlases were made in Portugal, Genoa, France, and Venice, and came in a variety of sizes from the smaller Italian texts through medium-sized Portuguese atlases to the grander Dieppe atlases. Illustration 52 is taken from a late-sixteenth-century atlas, most probably made in Genoa by Fransesco Ghisolfi. Portolan atlases were even made in the far-flung ports of empire. Fernao Vaz Dourado made seventeen chart portolan atlases in Goa in India.

By the middle of the sixteenth century, many portolan atlases were being produced. They differed in style and size and even in their coverage. The Portuguese-based atlases showed more detail of Brazil and East Asia, while French atlases revealed more of North America. In spite of the differences, there were many similarities. They were all hand-drawn on parchment and were beautifully decorated with elaborate gilt borders, delicate miniature paintings, and superb lettering. They were as much items of conspicuous consumption as records of European nautical voyages. They depict the world as a series of coasts and compass lines. At

52. Map of Eastern Mediterranean in Francesco Ghisolfi's *Portolan Atlas*, Genoa, ca. 1553; HM 28, ff. 10v–11r. This item is reproduced by permission of The Huntington Library, San Marino, California.

first glance some of the charts look like a word poem in the shape of a map. Coasts and ports are clearly depicted, continental interiors are hazy and often the setting for speculation and artistic license. The portolan atlases incorporate the knowledge of the mariner with the skill of the artist.

The Sea Atlas

Portolan atlases were luxury items. Handmade on expensive vellum, beautifully illustrated in the hand of gifted miniaturists, they were works of art meant for the rich and powerful. It is only by the end of the sixteenth century that we come across printed maritime atlases available for the mariner.

One of the first was *Spieghel der Zeevaerdt (Mariner's Mirrour),* published in 1584–85 with sea charts of the coastal waters of Western Europe from Cadiz to the Baltic. It represents one of the first printings of detailed sea charts. The author was Lucas Janszoon Waghenaer (1534/35–1606). He was born in Enkhuizen in the north of Holland. The town had grown rapidly in the sixteenth century, with its population growing from 3,500 in 1550 to 16,000 by the 1580s. The city was a vital trading link between the Baltic and Iberia; grain, herring, and wines flowed through the port. The town was directly involved in the great commercial expansion of the Dutch Republic. The citizens of the town invested 500,000 guilders, about 10 percent of the total capital invested, in the Dutch East India Company in 1602. The town was part of the great maritime tradition of the Netherlands that eventually would string a series of commercial connections around the world from Europe to the New World, Africa, and East Asia.

Waghenaer introduces himself in his book as a "simple citizen and pilot at sea." He went to sea very young, became a pilot, and came in contact with Portuguese, Spanish, and Italian navigators and mariners. By 1579 he was back on shore as an appointed tax official for the town, collecting excise monies, but he was dismissed from the post in 1582, allegedly for corruption. He then turned his attention full time to charts and maps. Although he had produced and sold three loose portolan charts and even made a map of the city, his first major work was *Spieghel der Zeevaerdt*. First published in Leiden in 1584–85, it is dedicated to Prince William of Orange. It provides a manual of navigation as well as printed charts on a common scale covering coasts of Northern and Western Europe. It also includes sailing directions.

The *Spieghel* is one of the first printed sea atlases. Waghenaer consciously modeled the atlas on Ortelius's *Theatrum*. Published in two parts, the first part appeared in 1584 (some contend it was actually published in 1583, although the title page indicates 1584); it begins with material on navigation along with tables of the sun's declination and lunar risings. It is a practical guide that gives detailed instructions on how to navigate by the stars. The frontispiece of the English translation depicts a variety of navigational instruments including a quadrant, cross staff, compasses, and, in a reference to the cosmographical discourse, celestial and terrestrial globes. The first volume contains a portolan chart of northwest Europe, the area covered in the atlas, followed by twenty-three detailed charts. The second volume, published the next year, contains twenty-two charts. All the charts are produced at a similar scale of 1: 370,000 and show profiles of the coast, compass roses, and depths around estuaries, along with the locations of sandbanks, towns, and rivers. Waghenaer brought together sea chart and land profile on the same page. The charts were finely detailed and the engraving were later used in book illustrations.

In 1584, Waghenear went to the trouble of getting a formal legal certificate drawn up with a sea captain and some first mates testifying to the authenticity of the charts. The captain found them (no doubt his good opinion was paid for) "good and useful that they should be printed for the benefit of seafarers."

The book was a great success. The first volume was reprinted four times in the first two years. The project had been a gamble; the atlas was expensive to produce, but Waghenaer found an eager market. Wealth and rehabilitation soon followed when the

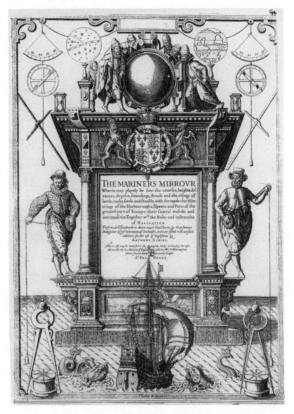

53. Frontispiece of *Mariners Mirrour,* London, 1588, an English translation of Janszoon Waghenaer's *Spieghel der Zeevaerdt* (1584–85). Courtesy of the Library of Congress, Washington, D.C.

city reappointed him to the excise office. Nothing succeeds like success.

The problem of language—few outside of Holland could speak or read Dutch—was quickly dealt with: a Latin edition was printed in 1586, making the book more accessible to a European intelligentsia. The two volumes were produced as one text. Vernacular editions from the Latin soon followed, with an English edition printed in 1588, a German version in 1589, and a French transla-

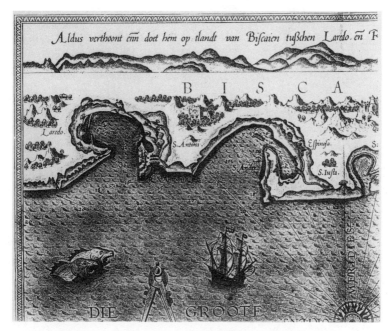

54. Detail from map in Lucas Janszoon Waghenaer's *Spieghel der Zeevaerdt*, Leiden, 1584–85. Courtesy of the Library of Congress.

tion in 1590. By 1592 it was available in five languages. Eighteen Dutch editions were published by 1620.

The success of the book lay in its innovation. There was nothing quite like it. Cornelius Koeman describes it as a "miraculous first."[4] It was a new and quite practical way to represent the world. There was a growing demand for this information; merchant shipping was increasing and there was always a demand from naval authorities. The English version was encouraged by Sir Christopher Hatton, Lord Chancellor and promoter of maritime commercial expansion. It was translated by Anthony Ashley, clerk to the Privy Council, who noted in the introduction, "Our most famous traveilours (to their endlesse renowne and honour of their countrie) have adventured the discoverie of diverse unknown coasts: and by the singular assistance of Almightie God, have compassed the Globe of the whole earth." The book became so popular that the term *waggoner* (after Waghenaer) became a common English word for a sea atlas.

Waghenaer went on to produce other navigational texts. His *Thresoor der Zeevaerdt* (Mariner's treasure) (1592) is a detailed set of sailing directions, more precise than *Spieghel* and covering more of northern Scotland, the Baltic, and the Mediterranean. Charts are at the uniform scale of 1: 600,000 with the profiles of coasts in the text rather than on the charts. *Enchuyser Zeecaert-Boecke* (1598) was a pilot's guide with no maps. Neither book was as lavish as *Spieghel*.

Waghenaer's work was the template for many subsequent publications. After 1590, as Dutch commercial expansion shifted from private ventures to a system of companies, there was a corresponding shift to organized, collated navigational information. In 1599, Willem Barentz produced an atlas of the Mediterranean. In 1606, Willem Blaeu (1571–1638) published *Licht der Zeevaerdt,* which updated *Spieghel* and included more charts of southeast England and the Low Countries. In 1612 this work was translated as *The Light of Navigation*. It was so successful when translated into English and French that it soon replaced *Spieghel* in the marketplace. Blaeu went on to publish *Eeste deel der Zeespiegel* in 1623; this book was translated two years later as *The Sea-Mirrour,* subtitled, "Containing a briefe instruction in the art of navigation and a description of seas and costs of the Easterne, Northerne and Westerne Navigation." The sea atlas concept that Waghenaer introduced was repeated down through the years. Amsterdam publishers were still producing maritime atlases as late as the end of the seventeenth century.

Navigation and Empire

The mapping of the oceans was intimately connected to the imperial ventures of European powers. Questions of navigation were not only applied cosmography—they were also applied politics. Hydrographical books, navigational treatises, and nautical charts were produced in a wider political and economic context. In the re-

Navigating the Seas 121

55. Frontispiece of Willem Blaeu's *The Light of Navigation*, Amsterdam, 1612. Courtesy of the Library of Congress, Washington, D.C.

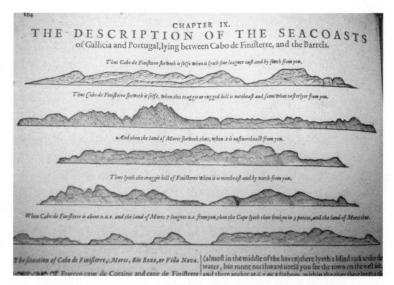

56. Coastal profiles in Willem Blaeu's *The Light of Navigation*, Amsterdam, 1612. Courtesy of the Library of Congress, Washington, D.C.

mainder of this chapter, I will limit my discussion to the early history of these texts in England.

The English were the European laggards of sixteenth-century overseas maritime expansion. The earliest printed European books on navigation were not produced in England. Giraldi's 1540 book, *De Re Nautica,* is a small book in Latin. It contains no diagrams and is meant for the scholar rather than the practicing seaman. More practical guides to navigation were provided by the Spanish, who regularly sailed across the Atlantic Ocean. Three important books were Pedro Medina's *Arte de Navegar* (1545), Martin Cortes's *Arte de Navegar* (1551), and Rodrigo Zamorano's *Compendio del Arte de Navegar* (1588). Medina's book was published in Valladolid, while the other two were produced in Seville. In his chapter titled "The Division of The Whole," Zamorano makes a distinction between theory and practice. As he writes, theory teaches us "the composition of the sphere of the world in general, and in particular of the heaven . . . and the circles which are imagined to be in that Sphere; without the knowledge of which, it is impossible to be a Navigator." Practice, on the other hand, involves the "use and making of instruments of navigation." These books contain diagrams and tables that aided practical mapmaking and navigation. They deal with the making of such instruments as the astrolabe, cross staff, and compass, and they include instructions on using these tools to draw a course. Full of diagrams and written in the vernacular, they are practical handbooks of navigational techniques.

The three books were published in a variety of languages. Efforts to publish them in English were not simply works of translation but conscious interventions into debates on national policy. Cortes's book was translated by Richard Eden, "Englyshed out of the Spanishe," as the subtitle noted. An ardent advocate of exploration and colonization, Eden (1521–70) was interested in cosmography and navigation. In 1553 he translated and printed "A treatyse of the newe India," part of the fifth book of Munster's *Cosmography,* which reported the voyages of Columbus, Vespucci,

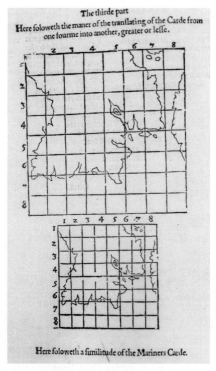

57. Page from *The Arte of Navigation*, London, 1584, Richard Eden's English translation of Martin Cortes's *Arte de Navegar* (1551). Courtesy of Dibner Library for the History of Science and Technology, Smithsonian Institution Libraries, Washington, D.C.

and Magellan. It was an early call to English colonization. His other translations included *A very necessary and profitable booke concerning navigation* and *A History of West Indies*. In *The history of trauvyle*, published in 1577, Eden translated Spanish works, including "varietie of matter, occurents out of forraigne countryes, newes of newe founde landes, the sundry sortes of gouernement, the differ-

58. Title page from *L'Arte del Navegar,* Venice, 1555, an Italian translation of Pedro Medina's *Arte de Navegar* (1545). Courtesy of Dibner Library for the History of Science and Technology, Smithsonian Institution Libraries, Washington, D.C.

ent manners and fashions of diuers nations, the wonderfull workes of nature, the sightes of straunge trees, fruites, foule and beastes and the infinite treasure of Pearle, Golde, Silver."

Eden's translation of Cortes's *Arte de Navegar* was published in 1561, with the express intent of "being published in our vulgar tongue, you may be assured to have more skilfull pilottes." An authoritative scholar of Elizabethan navigation noted that it was "one of the most decisive books ever printed in the English language. It held the key to the mastery of the sea."[5]

At mid-sixteenth century English maritime activity was confined to European waters. By century's end it had become globalized with the establishment of colonial settlements in North America and the Caribbean and trading posts strung across the globe in Africa, Asia, and the Americas. Navigational knowledge was an essential element in the creation and maintenance of this empire.

In the last quarter of the sixteenth century, there appeared an increasing number of English navigational texts, both translations and original works. A number of factors caused this growth. The English publishing industry was expanding. In 1510 only 67 books were published in England; this number increased to 92 in 1550 and then to 266 in 1600. Scholarship was also promoted and supported by rich patrons eager to promote their cause in print. The Earl of Leicester had 100 books dedicated to him, including the *Cosmographical Glasse*; Lord Burghley received 92 dedications; and Sir Walter Ralegh had 23. Often authors would dedicate their book to a powerful figure for their own advancement.

Writers who promoted improved navigation as a vital tool of English overseas expansion included our old friend John Dee. He brought back to England navigation and surveying instruments made in Europe. He was a technical adviser to the Muscovy Company who taught sea captains how to calculate latitude by accurate measurement of the ascension of sun and star positions. In his 1577 *Rare Memorials pertaining to the Perfect Arte of Navigation,* Dee suggested a large permanent fleet of sixty ships, wrote glowingly of a British empire, and encouraged English seamen to exploit the East. He also influenced a generation of scholars and writers, including William Bourne, Thomas Harriot, and Edward Wright. William Bourne, who wrote *A Regiment of the Sea* (1573), which includes navigational techniques and tables of declination, noted that he had visited the "great learned man" at Mortlake.

By the end of the sixteenth century the first full flowering of English navigational texts appeared. Thomas Blundeville, a Norfolk gentleman with an avid enthusiasm for applied cosmography, wrote a number of important texts. His *A Briefe description of universal mapps and cardes and their use* was published in 1589; five years later appeared his *Mr Blundeville His Exercises*. The subtitle noted that it was "necessarie to be read and learned by all young Gentlemen that are desirous to have knowledge as well as in Cosmographie, Astronomie and Geographie, as also in the Arte of Navigation." It was, as the introduction noted, "so plainlie written as any man of a mean capacitie may easilie learn the same without the helpe of any teacher." The book contained material on the principles of cosmography, a description of terrestrial and celestial globes, and a "new and necessary treatise of navigation" that discussed how to use the mariner's astrolabe, wind rose figures, and how to find the North Star and "the tydes in any place."

By the end of the sixteenth century, the English were also contributing to improvements in navigational techniques. A key figure was Edward Wright (1561–1615). He was born in Norfolk and educated at Caius College, Cambridge, in 1576. In 1589 he received permission from the Crown to go on an expedition to the Azores led by the Earl of Cumberland. The voyage honed Wright's interest in navigation. He returned to Cambridge in 1589 and worked on his book *Certaine Errors in Navigation*, which was eventually published in 1599. When his Cambridge fellowship expired in 1596, he was employed by rich merchants connected with trading companies to give mathematical lectures. The East India Company paid him a salary of £50. Wright was also the mathematics tutor to Prince Henry, the eldest son of King James. His book *Certaine Errors*, dedicated to Prince Henry, sought to improve navigation because, as he wrote, the field was "much stained with many blots and blemishes of error and imperfection." Sea charts were faulty, cross staff measurement was unreliable, and tables of star declensions were inaccurate. The first part of the book shows how to avoid the errors of the sea chart; the second deals with the complications caused by magnetic variation; the third takes up more

59. Title page of Edward Wright's *Certaine Errors in Navigation*, London, 1610. Courtesy of Dibner Library for the History of Science and Technology, Smithsonian Institution Libraries, Washington, D.C.

accurate use of the cross staff; and the final part discusses how to correct errors in the commonly used navigational tables. *Certaine Errors* became a hugely influential work that soon formed the basis for a more accurate English navigational practice.

Wright also corresponded with one of the most influential promoters of English overseas expansion, Richard Hakluyt (ca. 1552–1616). Hakluyt graduated from Christ Church College at Oxford in 1577. He took holy orders but was fascinated by the material world. In his public lectures at Oxford and in his letters and books, he continually enunciated the need to explore North America, to search for the Northwest Passage, and to establish overseas "plantations" that might stimulate national trade and wealth. In his *Discourse of Western Planting* (1584), he forcefully stated the economic benefits of overseas colonies and the need for state involvement.

Hakluyt's major publication was *The Principall Navigations, Voyages and Discoveries of the English Nation*. It first appeared as a one-volume edition in 1589 and was then published in an expanded three-volume version between 1598 and 1600. It is a work of economic nationalism: he wrote of his "ardent love of my country." A large, expensive book, it is composed of narratives and a long list of letters. The book is dedicated to the lord high admiral, the Earl of Nottingham. In the dedication Hakluyt mentions that he spent five years in France, where he "waded still further in the sweet studie of the historie of Cosmographie," a study that might "commend our nation for their high courage and singular activitie in the Search and Discoverie of the most unknown quarters of the world." The three volumes are an incredibly rich source of firsthand accounts of Elizabethan travel and discoveries. Written at a time of intense naval rivalry between England and Spain, the volumes are an extended praise of English exploration and discovery, a powerful boosterist text to promote further exploration and overseas expansion.

Hakluyt's work demonstrates how deeply the charting of the oceans was embedded in political and economic interests, allowing and encouraging the extension of trade routes and imperial ventures.

6

SURVEYING THE LAND

The term *survey* derives from the French *sur*, above or over, and the Latin *vigilare*, to watch. To survey is to watch over or watch from above. Thus a survey is an act of surveillance.

In the early part of the sixteenth century most land surveys were written records. Estate surveys, for example, consisted of written reports of the acreage of fields and manors, and, although detailed, they were written rather than mapped. Land was yet to be spatially coordinated, mapped, and plotted. By the end of the sixteenth century, however, surveys invariably involved maps, indeed the word *survey* began to mean a mapping exercise. Over the course of the sixteenth century, and especially during the second half of the century, surveying became a spatial practice.[1]

Surveying was first codified by Gemma Frisius (1508–55), already mentioned in chapter 2. Frisius revised and improved Apian's *Cosmographia*. He also enunciated the principles of triangulation in his book, *Libellus,* which in 1533 he had bound with the Flemish edition of Apian's *Cosmographia*. Triangulation involved surveying from different perspectives in order to build up a full picture of the terrain. If two points gave you a straight line, then three points gave you a survey. Frisius was an applied cosmographer with a knack for instrumentation and measurement. He outlined in principle how to calculate longitude through the transportation of time pieces.[2] He made globes and, in 1540, he published both a world map and an influential work on arithmetic that went through seventy-five editions over the course of the next one hundred years. In 1545, Frisius published a book on the cross staff; this work made the cross staff a more accurate tool for surveying. He knew John Dee, who visited him and introduced his work into England. The remarkable flowering of large-scale measured maps that appeared in Europe in the mid-1500s is in part due to Frisius, who had outlined both the principles and practice of mapmaking and surveying.

In the middle of the sixteenth century, specific books on surveying begin to appear more regularly. One example is Silvio Belli's small book on surveying, *Libro del misurar con la vista* (Book on

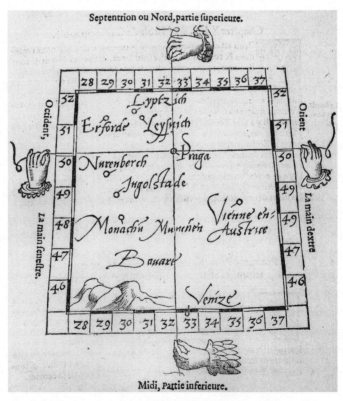

60. Gridded figure in Peter Apian's *Cosmographia*, Antwerp, 1581. Courtesy of Dibner Library for the History of Science and Technology, Smithsonian Institution Libraries, Washington, D.C.

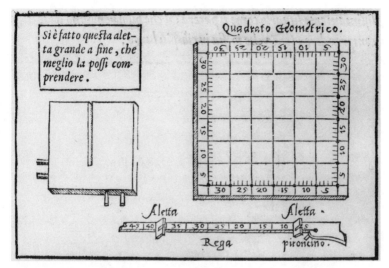

61. Illustration of surveyor's instruments and techniques in Silvio Belli's *Libro del misurar con la vista*, Venice, 1565. Courtesy of Dibner Library for the History of Science and Technology, Smithsonian Institution Libraries, Washington, D.C.

measuring by sight), published in 1565. It includes nine woodcut diagrams of elementary surveying tools and techniques. Surveying and surveying literature burgeoned in Europe in the latter half of the sixteenth century. One important reason was the growing commodification of land. Feudal lands were being turned into commercial lands. This increasing commercialization was reinforced by the price revolution of the sixteenth century. Beginning around 1470 and continuing to 1650, there was a steady increase in the cost of food, fuel, and shelter throughout Europe. The prime reason was population growth. In England, for example, population effectively doubled from 2 million to 4 million people over the course of the sixteenth century. Price increases matched population increases. The consequences included an increase in the price of land and the cost of capital along with a fall in wages. Societies became more stratified as the rich, especially those with land and capital, became wealthier and wage earners saw a decline in living standards. Land was becoming more valuable, increasing returns to landowners. Rent increases were higher than other price increases. Because land was worth more, there was an incentive for accurate surveying.[3] It is against this background of increasing land values that new surveying techniques were introduced and adopted. In this chapter, I will consider the development of surveying in England.

Estate Maps and Surveying in England

An important stimulus to surveying was the need of large landowners to have an inventory of their holdings. From the tenth century to the sixteenth century, English estate records were in the form of written descriptions. Even in the middle of the sixteenth century estate maps were rare, but by the end of the century they were commonplace. Surveys became both spatial practices as well as written reports.

The surveying of estates was intimately connected to the growing commercialization of the land market. Surveying as a practice and as a profession developed in the fluid and appreciating land market in the wake of dissolution, enclosure, increasing valuation, and ongoing disputes and litigation.

The commodification of land in England was fueled by the dissolution of the monasteries in the 1530s and the enclosure movement. Henry VIII's break with Rome involved the destruction of the land holdings of the Catholic Church. There was a property bonanza as monastery lands were turned into private holdings. More land was coming into the pool of private land that needed to be itemized and surveyed. In the county of Wiltshire, for example, the land of the monastic orders was taken as private estates and manors by landowners: the Earl of Pembroke acquired Wilton, Sir John Thynne got Longleat, and Sir William Sharington added Lacock with money stolen from the mint. The clothier William Stumpe turned part of Malmesbury Abbey into a cloth factory and used the other part as his private residence.

The enclosure movement was essentially the privatization of common land into private land. The common land used by the peasantry was turned into private holdings, especially holding of the largest landowners. The result was the commodification of land and increasing inequality as the landowners became richer and many villagers became landless peasants.

Some of the earliest estate maps were more pictures than maps. The delightful map-picture of Wotton Underwood, done at some point between 1564 and 1586, is a good example of the pictorial estate map. But as the land market heated up in the sixteenth century, more accurate surveys were needed and more surveying texts began to appear.

Surveying drew upon a basis of mathematical competency. An important figure was Leonard Digges (ca. 1520–59), who had studied with John Dee and also traveled abroad, where he became more familiar with the work of Peter Apian and Gemma Frisius. In

62. Pictorial estate map of Wotton Underwood, ca. 1564–86; HM 26. This item is reproduced by permission of The Huntington Library, San Marino, California.

1556, Digges published an elementary surveying book, *Tectonicon*, which described the instruments needed for survey and discussed methods of survey calculation. His work was developed further by his son, Thomas Digges, who was born in Kent around 1546; Thomas received mathematical training from both his father and John Dee.[4] Digges the son completed the book *A Geometrical Practise, named Pantometria* that his father had begun. Published in 1571, the book is divided into three parts. The first, "Longimetra," concerns the measurement of length, height, distance, and depth.

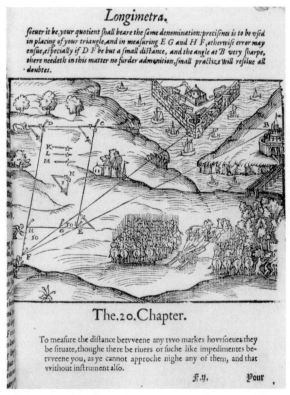

63. Page from Thomas Digges's *A Geometrical Practise, named Pantometria*, London, 1571. Courtesy of Dibner Library for the History of Science and Technology, Smithsonian Institution Libraries, Washington, D.C.

The second part, "Planimetra," concerns the measurement of area. It begins with a simple geometry of triangles and bounded objects before going on to show "Howe you maye from an highe Hil or Cliffe, measure how manye Acres, Roodes or Perches, is contayned in any Fielde, parke, Wood or other playne Superficies in the countrie rounde aboute you." Part three, "Stereometria," deals with the measurement of solid objects. The work presents knowledge as being an aid to making a living in the world; it is mathematics applied to the mercantile world.

While the Digges's book deals with surveying as a form of applied mathematics, Valentine Leigh's *The moste profitable and commendable science* (1577) concentrates exclusively on land surveying. It is a small book on cheap paper meant for a wider readership. The emphasis is on "how to make a perfecte surveye to moste profite." Leigh's book notes that the surveyor should know more than how to make a plot of the whole manor; he should also know and consider all rents and know the law as it applies to estates and manors. Surveying texts throughout much of the early sixteenth century were of two types: instruction books for land stewards and measuring manuals. Respective early examples are *The Boke of Surveyeng* by John Fitzherbert (1523) and *The Maner of Measurying* by Richard Benese (1537). Valentine's book combines both types, being equally concerned with actual surveying techniques and written reports. The book is a guide to spatial surveillance for the land managers of the new manor class.

Surveying as a spatial practice became more common during the final quarter of the sixteenth century. Edward Worsop's *A Discoverie of Sundrie Errours and Faults daily committed by Landemeaters, Ignorant of Arithmetike and Geometrie* was published in 1582. Dedicated to Lord Burghley, the book notes that rents are determined by acreage so that it is vitally important to estimate the correct acreage with accurate measurement. Edward Worsop was one of the first to suggest in print that surveyors draw maps.

Ralph Agas's 1595 *A Preparative to Platting* is a small pamphlet of only twenty pages that promotes the spatial practice of surveying. He describes in detail how to survey an estate and write up the survey. The same techniques, he notes, can be used for "platting a countie or shire, or Citie, Borough or Towne." Agas argues the need for better measurement with accurate instruments; "You shall first have a plain table, faire spread with white cover... There is also the Staffe, astrolabe, square, ring ruler, circumferentor, sector and half protractor and theolodite." Agas gives very practical advice and provides examples of the costs of mistakes caused by inaccurate surveys; in one case he shows how an improper survey could devalue land from £9 per acre to 10 shillings per acre. Mistakes in surveying could be expensive.

There was a close connection between mapmaking and surveying. The mapmaker Christopher Saxton was also a surveyor. After completing his atlas of England he was fully employed as an estate surveyor from 1587 to 1608, completing about four surveys each year. He made around twenty-five estate maps and fourteen written surveys. Saxton made a variety of survey maps to resolve boundary disputes and sort out competing claims on water rights. He also created general estate maps such as the one he did for Michael Wentworth, who had purchased estates in Yorkshire in 1599 and wanted a map to better know his new lands and tenants.

Estate maps were often very detailed. Ralph Agas's 1581 manuscript map of Lord Cheney's estate of Toddington, for example, is to a scale of 40 inches to one mile. The final map, which comprised twenty parchment sheets measuring 11 feet by 8 feet, allowed Lord Cheney to see his whole estate in a single viewing. While involved in the technical side of surveying, Agas knew that increasing rents in the wake of new surveys could involve "a dangerous harming of peace betweene the Lord and his tenants between neighbor and neighbor."

The social consequences of surveying were developed more fully by John Norden (1548–1625), a practicing surveyor and mapmaker. His work covers three important cartographic endeavors of Elizabethan-Jacobean mapmaking: county maps, city maps, and estate maps. Norden worked in all three areas. He had ambitious plan to produce a series of county studies including maps. Only two were published during his lifetime: Middlesex in 1593 and Hertfordshire in 1598. He produced manuscript maps of Essex, Northampton, Cornwall, Kent, and Surrey. His work provides the basis for some of John Speed's atlas, *The Theatre of the Empire of Great Britaine,* and William Camden's 1607 *Britannia.* Norden was something of an innovator; he devised a grid and gazetteer so that places could be identified, he was one of the first to include roads on his maps, and he devised a triangular distance table.

Norden also produced a map of London in his Middlesex study as well as views of London in 1604–6. Norden's map of Shakespearean London, situated between the more celebrated Agas woodcut of ca. 1560 and the Ogilby and Morgan map of 1676, is an important example of early urban mapmaking.

Norden was also a surveyor of estates. As the surveyor of Crown Woods, his official title was General Surveyor of Kings Landes and Woods; he was also surveyor to the Duchy of Corn-

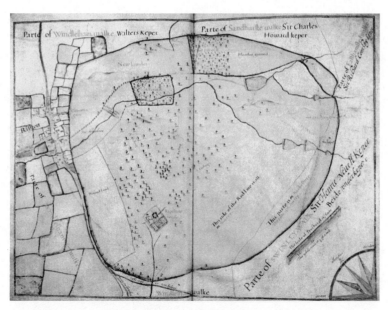

64. Map of Bagshot Park in John Norden's *A Description of the Honor of Windsor,* 1607; K38669 ff.17. By permission of the British Library, London.

wall. The British Library has a rich collection of his manuscript maps. It also holds Norden's *A Description of the Honor of Windsor* (1607), an atlas of estate maps of the royal holdings in Berkshire, Surrey, and Buckinghamshire; it is dedicated to Prince Henry. The Royal Library at Windsor Castle has another copy dedicated to King James. The atlas contains more than seventeen manuscript maps of individual deer parks giving the general layout as well as the keeper's name and number of deer. Illustration 64 shows Bagshot Park, which contained seventeen "Rowe" deer.

Norden's estate maps are especially interesting when read alongside his book of 1607, *Surveyors Dialogue*. While the maps embody the enclosure movement and the power of the landed gentry and royalty, the book is acutely aware of the social relations and conflict involved with the enclosures and subsequent surveying. The book is a technical document of how to survey that is also deeply aware of the new social relations emerging from the commodification of land and the growing tension between social obligations and monetary concerns.

The subtitle promises Norden's book to be "very profitable for all men to peruse but especially for all Gentlemen or any other Farmer or Husbandman that shall have occasion, or be willing to buy or sell land." In the preface Norden begins with divine praise of the earth as a gift from God, but he notes that it is up to man to improve upon this earthly revenue. He argues that a balance needs to be struck between increasing rents too much, which "will afflict the hearts of poor tenants," and allowing rents that are too low, called "an absurd leniency that breeds the contempt of the tenants." The first part of Norden's text is in the form of a dialogue between a farmer and a surveyor, "wherein it is proved that surveys are necessary and profitable for both lord and tenant." The farmer says, "this is an upstart art found out of late, both measuring and plotting." The surveyor tells the farmer that land value has increased, thus requiring more accurate surveys. The farmer argues that surveys are the cause of increased rents, but the surveyor replies that it is important to know the truth of the land; "is it not lawfull for the Lord of the Manor to examine of the things belonging unto him." The second part of the text is a dialogue between the lord of the manor and the surveyor. The surveyor reminds the lord of his obligations to his tenants. Norden puts the surveyor in the middle ground between increasing the "Lord's profit while still maintaining the obedience and favor of the tenants." In Norden's text, surveying is less a simple technical exercise than it is a complex balancing act between increasing rents and maintain-

65. Title page from Aaron Rathborne's *The Surveyor*, London, 1616. Courtesy of Dibner Library for the History of Science and Technology, Smithsonian Institution Libraries, Washington, D.C.

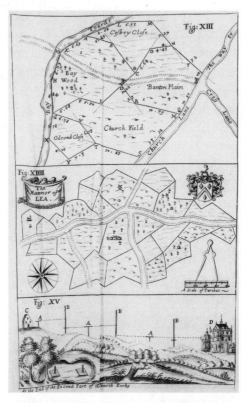

66. Diagrams in William Leybourn's *The Compleat Surveyor*, London, 1674. Courtesy of Dibner Library for the History of Science and Technology, Smithsonian Institution Libraries, Washington, D.C.

ing social harmony. Norden's book is both a technical handbook and a moral tract.

Norden's book is perhaps the last surveying text to be so socially sensitive; later books had little awareness of the social context and were more technical texts. Aaron Rathborne's *The Surveyor* (1616) is one of the most comprehensive texts of the early English surveying books. Dedicated to the Prince of Wales, the book is in four parts. The first two parts deal with principles of geometry, while the last two cover the "matter of survey." The last two books derive from what Rathborne sees as the abuse of surveying tools by simple folks. He was in effect creating a professional discourse of surveying. The book was still being used as a primer almost two hundred years later. But there is no mention of the social obligations of the landowner or the rights of tenants. The surveyor's gaze looks out on a land that is completely commodified; the gaze is one of measurement only for the purpose of profit.

Rathborne's book was cosmography applied to the new professional discourse of surveying. Land surveys were part of the enclosure of common land into private hands, the commodification of land, the increasing commercialization of land; they served as the basis for the renegotiation of leases and rent levels. But while Norden was aware of the social consequences of land surveying, Rathborne takes it for granted. After Norden, surveying soon become an accepted technique of landlord surveillance. By the late seventeenth century, surveying was a professional discourse concerned with precision and accuracy. The commodified, surveyed land revealed in illustration 66 (although this book was published in the late seventeenth century) is an example of the new spatial practice that emerged in the sixteenth century. The practice of surveying had broken off from a general cosmography to become a technical, professional matter, abstracted from social context, used primarily by the rich to measure and assess the value of their landholdings.

7

ANNEXING TERRITORIES

Between July 1585 and June 1586, two Englishmen traversed the outer banks and coasts of what was subsequently called Virginia. One had an artist's notebook, the other had a plane table and series of surveying instruments. Traveling by boat, the two men recorded and painted, wrote and surveyed. They made a map, wrote a report, and created an artistic rendering of all that they saw. Together they made one of the most accurate early maps of North America. One of the men was Thomas Harriot; the other was John White.

Colonial Cartographies

Thomas Harriot (1560–1621) was a friend and student of John Dee. He was a cosmographer with specific interests in mathematics, surveying, and navigation. He popularized the use of the equal sign (=) and invented the signs < and > to refer respectively to less than and greater than. Soon after graduating from Oxford University in 1590, he was employed by Sir Walter Raleigh to teach navigation. In March 1584, Raleigh received permission from the government to make voyages to the New World. Later that year, the first expedition, led by captains Amadas and Barlow, made it across the Atlantic. Harriot drew up the navigational instructions for the two captains. Harriot was an important member of the second expedition, composed of seven ships, which left Plymouth on April 9, 1585. For this second voyage Raleigh decided that White and Harriot should go; he instructed them to note down and draw what they saw, survey the ground, and make a general map. Harriot took a variety of instruments with him including compasses, clocks, universal dial, cross staff, back staff, and almanac tables. He was the principal navigator on the voyage. The fleet first landed on a small island of Puerto Rico and later, on June 30, the fleet anchored at Roanoke Island.

John White was a painter and draftsman. Born around 1540–50 he became an active participant in England's overseas exploration. He went on Martin Frobisher's 1577 expedition into Arctic waters and made some of the first and most detailed depictions of Inuit peoples. Before the advent of photography, artists were important members of expeditions, recording sights and images to be taken back home. White went on five different overseas explorations.

Soon after landing, Harriot and White began their survey of the New World. Beginning around July 11, 1585, Harriot and White moved up and down the Carolina outer banks by boat and surveyed the entire coastline from Ocracoke to Cape Henry; they also went west up the Roanoke and Chowan rivers. They spent almost a year, using their cosmographical knowledge and artistic skills to record and plot. Harriot fixed exact locations with his astronomical instruments and survey equipment and made observations about the local people and their language. White documented the scenes they witnessed. He painted sixty-three watercolors of the local flora and fauna as well as scenes from the everyday lives of the native peoples. The result was one of the first comprehensive surveys of North America by Europeans.

Both men returned to England in 1586. Harriot wrote *A Briefe and True Report of the New Found Land of Virginia* for popular consumption, meant to entice and attract as well as record and document. The book, published in 1588 with no illustrations, was divided into three parts: discussions of "commodities that can be traded," such as silk grass, flax hemp, wine, and dyes; "commodities that provide sustenance," such as walnuts and medlars; and "other commodities." There was a large section on the nature and manners of the local people. It was essentially a depiction of a territory ripe for exploration and appropriation.

There was tremendous interest in American voyages. The Flemish publisher Theodor de Bry had plans to publish a multilanguage American voyage series. Harriot's book, republished as the first in this series, appeared in 1590 in Latin, French, and German. It was larger than Harriot's original report with the addition of illustrated engravings from John White's watercolors. In a section titled "The True Pictures and Fashions of the People in That Parte of America now called Virginia, discovered by Englishmen," Bry transformed twenty-three of White's drawings into elegant engravings, providing detailed documentary evidence of life at first contact. These illustrations provide one of the most detailed early European responses to North America. The book was enormously successful and the images have influenced European perceptions of Native Americans to the present.

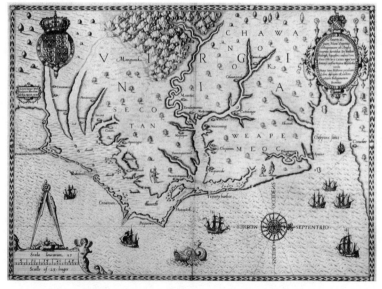

67. Map of Virginia in Theodor de Bry's *Admiranda narratio,* Frankfurt, 1590; RB 72400. This item is reproduced by permission of The Huntington Library, San Marino, California.

One of the plates, taken from an original manuscript, is a very detailed map of the outer banks. Drawn at an accurate scale of 25 leagues to the inch, it shows a carefully delineated coastline and the site of Native American villages. The map is the result of the careful survey done by the two men in 1585–86. This map, drawn from accurate astronomical observations and employing a uniform scale, was a combination of Harriot's mathematical ability and White's artistic sensibility. There is another smaller map, accompanying a section titled "The arrival of the Englishmen in Virginia," which shows in greater detail the location of Roanoke. The maps represent many things in addition to the coastline of a new territory. They signify the spatial practice of an applied cosmography brought into the service of the colonial enterprise.

This particular enterprise was not successful. Internal dissension combined with starvation and disputes with the local people,

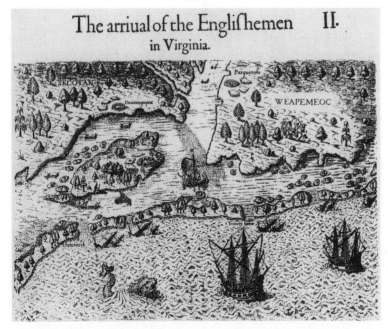

68. Detailed map of Virginia in Theodor de Bry's *Admiranda narratio*, Frankfurt, 1590. Courtesy of the Library of Congress.

whose food stores had been robbed, led to the first permanent English settlement in North America being abandoned in June 1586. Many of White's drawings were lost during the hurried departure in rough seas.

Naming John White as governor, Raleigh sent out yet another expedition in 1587 to establish the "Cittie of Raleigh." This colonial venture, which brought 110 people back to Roanoke, was no more successful than the previous one. Hunger was a constant companion and White went back to England for more supplies. His return was delayed, and when he finally made it back to Roanoke in 1590 he found the colony abandoned. White noted that his chest had been ransacked, "with the frames of some of my pictures and Mappes rotten and spoyled." In a letter he sent to Richard Hakluyt on February 4, 1593, from his house in Newtown in Kilmore (the gift of a grateful Raleigh), White wrote of his Virginia voyages as "luckless to many as it was evil to myself."

When Harriot returned to England in 1586 he was again employed by Sir Walter Raleigh, who gave him an estate in Ireland. Harriot teamed up again with White, and together they surveyed and mapped Raleigh's Irish landholdings. When Raleigh's influence began to wane, Harriot was fortunate to find another generous patron, the ninth Earl of Northumberland, who gave Harriot accommodation at Syon House in London and funded his inquiries. Here Harriot continued to work on astronomy, optics, alchemy, and mathematics; he even designed a new water system for Syon House. He died in 1621 after a long and painful illness. Smoking, a habit he picked up in Virginia, may have been a major cause. He was, perhaps, one of the first European casualties of heavy tobacco use.

Colonial Mappings: Center and Periphery in Cartographic Encounter

Philip II of Spain was the ruler of the first global empire. In 1556 he succeeded to his father's dominions in America and most of Europe. At its height, his empire stretched around the globe, encompassing not only much of Europe including Spain, Naples, the Netherlands, and Portugal, but also New Spain in present-day Mexico, Peru, Brazil, and the spice islands of Southeast Asia.

Philip, who ruled from 1556 to 1598, was a typical Renaissance king, much taken with the new cartography as a way to see and embody his territories. He set about to systematically collect spatial data about his empire. In 1559 he set up a "general Visitation" for each of his dominions in Italy. He also employed Anton van den Wyngaerde to prepare a series of Spanish cityscapes; sixty-two views of fifty cities remain. He commissioned a complete map of the Iberian peninsula from the cartographer Pedro de Esquivel. Work began in the 1560s and 500,000 square kilometers were surveyed; the results were portrayed in the twenty-one-sheet Escorial Atlas. Mapped at the detailed scale of 1: 430,000, it was then the most detailed survey of mainland Europe.

The mapping of territory was an important element of Philip's maintenance of knowledge and power. As his nation and empire expanded, Philip was unable to visit all of his empire. He spent more time at the imperial center, the Escorial Palace near Madrid, as he got older. Maps were a way of seeing, a technique for surveying and visualizing and hence controlling his kingdom. Skilled cartographers were employed to make large-scale maps and detailed topographies of individual cities.

Questionnaires were also regularly sent out from the center to the rest of Spain and the overseas empire. Returns for more than 600 villages in Castille have survived. The royal cosmographer, Juan Lopez de Velasco, was also commissioned to compile two works: *A geography and general description of the Indies* and *A demarcation and division of the Indies*. The two works were presented to the council of the Indies in 1574–75. In 1577, Lopez sent out a three-page, fifty-item questionnaire to officials in the New World under Spanish control, requesting them to provide written and cartographic information. The aim was to obtain information on latitude and longitude, written commentaries, and accurate maps. The replies from the New World are called the *Relaciones Geográficas* (Geographical reports).

148   ✤   *Making Space*

The survey was not a great success. Some questions were rarely answered and the maps were not drawn to a systematic and coherent cartographic convention. However, the survey is enormously revealing—less for the imperial Spanish goal of creating an accurate map of the New World and more for providing an example of a cartographic encounter.

Part of question 10 of the original questionnaire requested, "Make a map of the layout of the town, its streets, plazas and other features." In the *gobierno* of New Spain, an area similar to contemporary Mexico, ninety-eight responses were received.[1] The written comments were made by the local administrators, but the

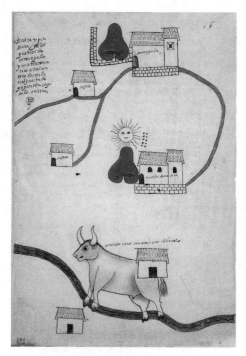

69. The *Relacione Geográfica* map of Tenanpulco and Matlactonatico, 1581. Photography courtesy of the Benson Latin American Collection, University of Texas at Austin.

maps were drawn by a variety of people. Two-thirds of them were drawn by indigenous people. When told by the Spanish officials to answer the cartographic query posed in question 10, the indigenous cartographers responded by painting a map of the community. Their maps are less Spanish town plans and more examples of indigenous cartographic tradition that combined pre-Hispanic elements to depict places, people, boundaries, and architecture. The mapmakers drew upon a rich tradition that expressed community identity, history, and the social order. Another level of meaning was added as the Spaniards wrote on the maps; the inscriptions told of place names, the hierarchy of settlements, and ways the local areas were tied to a broader Spanish empire. The indigenous maps were thus inscribed by the notations of imperial dominance. One Spanish scribe, for example, wrote on the routes out from a city, indicating where they led to in the Spanish Empire.

The maps of the *Relaciones Geográficas* show, in cartographic form, the interplay between political imposition from the center and local response: how the discipline of the center became a practice in the periphery; how a top-down exercise became a bottom-up response; how center and periphery coalesced in a transcultural, cartographic encounter. By depicting indigenous spatial representation, the maps give vivid testimony to the power of mapping and the mapping of power in one colonial moment when center and periphery were connected in a transaction of cartographic production and consumption.

8

CONCLUSIONS

Space

Space is both an absolute and a relative. It is an ontological necessity, the basic container of material substance, as well as a relative concept, a social construction, because how we see space is not predetermined. Space is perceived as well as inhabited.

During the sixteenth century space as a social construction was reimagined. It was revisioned and reconceptualized from a locus of religious wonder to an instrumentalized sphere marked for exploration, conquest, and appropriation. This shift involved the revival of the classical authors and new forms of instrumentation.

The work of Ptolemy and Euclid was extended by a range of commentators eager to fashion a new world on the foundations of classical authorship. From its widespread appearance in the late fifteenth century, Ptolemy's *Geography* was a major influence on the construction of a more precise geographical discourse. Euclidean geometry furnished a language to describe space that transcended the particularities of time and place. Location could be registered, with reference to latitude and longitude, in a numerical grid system. The world was perceived and understood through this grid of latitude and longitude, replacing history with geography and time with space. Some people in different parts of the world were imprisoned in and by the grid. The grid itself was twisted and turned in a wide array of map projections. In the cities of Europe, cosmographers experimented with new ways to represent the world: the cordiform projection was first used in 1500, the simple conic in 1506, the azimuthal equidistant projection in 1518, and the azimuthal orthographic projection in 1529. Mercator's projection was developed first in 1569 and refined in 1599. The sixteenth century was one of the most creative periods in the history of cartographic projections.

New instruments and techniques, such as triangulation, were also developed to accurately calibrate this grid. A new form of space—a measured, surveyed space—was created by this instrumentation of the world. The world was seen through the eyepieces of optical instruments and recorded in mathematical notation in an unblinking gaze that cataloged and gridded.

The major forces driving this revisioning were a number of commercial and national interests. The commodification of land, the annexation of colonial territory, the navigational needs of long-distance merchant shipping, and the surveillance requirements of national governments and state authorities both prompted improvements in mapping, surveying, and navigation and created a cartographic language that envisioned—even while it appropriated—the world.

Cosmography

Central to many aspects of European intellectual, social, and political life in the sixteenth century, cosmography was an inclusive spatial practice whose goal was a comprehensive understanding of the cosmos, both celestial and terrestrial. It drew upon classical and medieval scholarship in a single encompassing sweep of the world with the sphere—a symbol of order, perfection, and unity—as its recurring motif. At the beginning of the sixteenth century, cosmography maintained its coherence, but by the end of the century it was fractured into separate subjects. A comprehensive cosmography gave way to distinct fields of spatial practice of astronomy, geography, surveying, and navigation, which remain recognizable today.

The story of cosmography also tells a bigger tale of the shifting nature of organized knowledge, the changing nature of knowl-

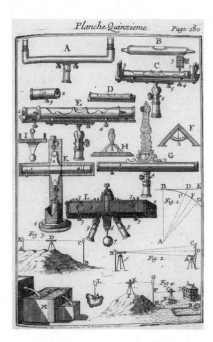

70. Surveying instruments in Nicholas Bion's *Traite de la construction et des principaux usages des instruments de mathematique,* Paris, 1709. Courtesy of Dibner Library for the History of Science and Technology, Smithsonian Institution Libraries, Washington, D.C.

71. Mapping instruments in Nicholas Bion's *Traite de la construction et des principaux usages des instruments de mathematique,* Paris, 1709. Courtesy of Dibner Library for the History of Science and Technology, Smithsonian Institution Libraries, Washington, D.C.

edge formation and epistemological reconstitution. This reordering of what constitutes knowledge, along with the resultant divisions and demarcations, influences present-day understanding of the world around us. The rupturing of astronomy from astrology and science from magic in the fragmentation of the coherent cosmography of the sixteenth century, for example, marks a division between "science" and "nonscience," giving us a very different worldview.

There is something deeply appealing about the cosmographical project. It was the scholar's fantasy of cerebral mastery of the world. Alas, the division of knowledge has now balkanized the academy, making us perhaps deeper but narrower in our understanding. But there is still a dream that lies deep in the heart of many writers and scholars—that they can encompass (a telling metaphor) the world within a conceptual framework, that the world is subject to a complete understanding in one totalizing sweep. The search for a general theory of the world counterpoised with the discovery of the endless intricacies of our universe provides a hugely creative tension in the history of human understanding of the world.

Envisioning the World

The fragmentation of cosmography was not simply a reordering of knowledge; it was an epistemological rupture that involved the development of new spatial practices, astronomy, geography, topography, navigation, and surveying. Exploration, discovery, the rise of a mercantile class, the creation of a commodified land market, the cartographic display of national territory, and the annexation of colonial territory ingrained and refined these discourses. In their turn, these discourses generated more sophisticated forms of commodification, nationalism, and colonialism, ultimately aiding in the appropriation of space for material ends. Spatial practices were the bridge between an understanding of the world and a transformation of the world.

The spatial revisioning had distinct textual forms. At the intersection of reading and writing, new textual forms of space were created in the sixteenth century: the first modern printed atlas (Ortelius's *Theatrum*, 1570); the first printed urban atlas (Braun and Hogenberg's *Civitates*, 1572); the first printed national atlas (Saxton's *Atlas*, 1579); and the first printed maritime atlas (Waghenaer's *Mirrour*, 1584–85). The world was seen and represented as text. The world was encompassed in texts.

Although I have been using the term "revisioning," a more accurate term would be "envisioning." It was not so much that a view was changed but that a view was created. The Renaissance privileged the eye with its emphasis on spatial form rather than enduring essence. This ocular fascination is apparent in the titles of some of the many books I have considered: Cuningham's *Cosmographical Glasse* (1559), Waghenaer's *Mariners Mirrour* (1584–85), Blaeu's *Light of Navigation* (1606), and John Norden's proposed *Speculum Britanniae* (Mirror of Britain). Glass, mirrors, and lights all appeal to a concern with seeing, vision, and spatial form.

The spatial envisioning is also apparent in the notion that developed in the sixteenth century that the world was a stage. It was only in the mid-sixteenth century that the term theater was used, but Abraham Ortelius's *Theatrum Orbis Terrarum* of 1570 was just one of many cosmographical/geographical books that used the term "theater" as a spatial signifier; other examples included Giovanni Paolo Gallucci's (1589) *Theatrum Mundi* and John Speed's (1611–12) *Theatre of the Empire of Great Britaine*. By the end of the century the term was so commonplace that it could be used in contemporary drama. "All the world's a stage," wrote William Shakespeare in his 1599 *As You Like It*, "and all the men and women merely players."

The relationships between drama and cosmography are many. The globe was not a only a tool but also a metaphor. Shakespeare's plays were performed, let us remember, at The Globe. The notion that the world was a setting for human action that could be both visualized and analyzed was just one of the links that bound the new spatial practices with the new forms of dramatic enactment.

The globe is also a visual representation of the world that reinforces the notion of one world. There is a connection between the development of the globe and the sense of one complete and connected world. "Globalization" and "globe" share the same root. The more widespread use of the globes in the sixteenth century reflects both new forms of spatial instrumentation and a new spatial sensitivity. The globe embodied a new form of globalism based on increased geographical knowledge, global mercantile trade, and global colonial appropriation.

Magic and Mystery

The new spatial envisioning in the sixteenth century was not a proto-science emerging from medieval murkiness. It was still deeply marked by magic and religion. The iconic figure of John Dee, mathematician, mapmaker, communicator with angels, and sometime alchemist, embodies the paradoxes of the age. The sixteenth-century cosmographers firmly believed in astrology and alchemy as well as astronomy and mathematics. Navigation and prognostication went hand in hand. There was no rift between the new spatial discourses and religious beliefs. The cosmographers saw their work as revealing God's wonder; to map the world was to reveal the "fabick of his creation." The cosmography of the sixteenth century cast a wide arc in a complex hybridity of magic and science, belief and measurement, rational narrative and religious iconography.

Spatiality Today

The spatial envisioning of the sixteenth century held sway for almost five hundred years. It is fitting, then, to end with the tantalizing notion that the early twenty-first century may be witnessing a spatial reinvention as radical as that of the sixteenth century. Since the 1960s images from space have given us a new global perspective, and since the 1970s satellites have provided real-time global coverage. We can watch the ozone hole above the Antarctic wax and wane from satellite imagery. We can now see the drama of forest fires in Southeast Asia, the urbanization of the Chinese countryside, and the urban sprawl of the continental United States unfold across our computer screens.

The coordinates of the new electronic space mark a still-undefined territory that is part cyberpunk matrix, part digitized archive—a realm between myth and science similar in many ways to the cosmography of the early Renaissance. The increasing instrumentalizing of cyberspace seems to facilitate westernization just as its early modern counterpart did; this time, however, it is not in the form of colonial expansion but rather in the trajectory of market penetration. Yet similarities between spatiality in the age of exploration and spatiality in the age of information should not be forced. The unblinking, unopposed gaze that once charted the world in terms of the grid is now returned, engaged, refracted, refused, avoided, and occluded in the interactivity of cyberspace. The grid of latitude and longitude may not imprison in a world that privileges the nonlinear, the fluid, the peripheral. Maybe?

Let us return again to the image at the beginning of this book, the portrait titled *Cardinal Bandinello Sauli, His Secretary and Two Geographers*, painted by Sebastiano del Piombo in 1516. Almost five hundred years later, what would the comparable image be? Most likely not an oil painting—more likely a music video. And while cardinals still wield some power, the main figure would now be a CEO of a global company. And the book that the two geographers were looking at in the Piombo painting would be replaced with multiple screens showing feeds of news from around the world, market reports, and a variety of real-time images from surveillance cameras and satellites. The contemporary equivalents of Piombo's two geographers would be computer hackers hunched over the screens, tapping keys, accessing and displaying information. They would be looking at and interacting with a world represented in real-time, multiple, blinking screens.

APPENDIX
NOTES
SELECTED BIBLIOGRAPHY
INDEX

APPENDIX

A Selection of the Renaissance Translations of Ptolemy's *Geography*, 1475–1600

Date	Maps	Text	Printed
1475	0 maps	Latin	Vicenza[a]
1477	27 maps	Latin	Rome[b]
ca. 1480	31 maps	Italian	Florence[c]
1482	32 maps	Latin	Ulm[d]
1507	33 maps	Latin	Rome[e]
1511	28 maps	Latin	Venice[f]
1512	1 world map	Latin	Cracow[g]
1513	47 maps	Latin	Strasbourg[h]
1514	no maps	Latin	Nuremberg[i]
1540	48 maps	Latin	Basle[j]
1548	60 maps	Italian	Venice[k]
1561	64 maps	Italian	Venice[l]
1578	28 maps	Latin	Cologne[m]
1596	64 maps	Latin	Venice[n]
1597	10 maps	Latin	Louvain[o]

[a] Based on the 1409 Latin translation from Greek by Jacopo d'Angiolo.

[b] Amended version of d'Angiolo translation. The maps are perhaps the first examples of copperplate engravings for books. Reprinted in 1490.

[c] Metrical paraphrase by Francesco Berlinghieri. Reissued in 1500.

[d] The d'Angiolo translation amended by Donnus Nicolaus Germanus. The maps are woodcuts and include five contemporary maps. Reprinted in 1486.

[e] A new edition of the d'Angiolo translation amended and revised. Copperplate maps include six new modern maps. Reissued in 1508 with a modern map of

the world by Johann Ruysch; this is the first printed map to contain the New World.

ᶠ The d'Angiolo translation edited by Bernardus Sylvanus. Woodblock maps include a cordiform (heart-shaped) projection of the world with the first printed delineation of North America.

ᵍ Introduction to Ptolemy's *Geography* with a woodcut map of the world in two sheets. Second edition printed in 1519.

ʰ Begun by Martin Waldseemuller. The woodcut maps include twenty contemporary maps and one of the first maps devoted to the New World, "Tabula Terre Nove." Reprinted with minor changes including maps bearing the name America in 1520, 1522, and 1525. Laurens Fries edited the third edition of 1522.

ⁱ A new translations with additional geographical works by other authors.

ʲ A new edition revised and edited by Sebastian Munster, who designed the forty-eight woodcut maps. Reprinted in 1542, 1545, and 1552.

ᵏ Italian translation by Pietro Andrea Mattioli of Siena. The copperplate maps, drawing on Munster's maps, were designed by Jacopo Gastaldi.

ˡ New Italian edition with new maps. Reprinted in 1562, 1564, 1574, and 1598.

ᵐ This volume contains twenty-eight copperplate maps by Mercator. No text. Published in 1584 with text.

ⁿ A new Latin edition with new maps by Giovanni Antonio Magini of Padua. Subsequent editions were published in Cologne and Venice in 1597.

ᵒ Wytfliet's supplement to Ptolemy's *Geography*. The text and maps refer to America; probably the first American atlas. Second edition published in 1598.

NOTES

Epigraphs

1. Edward Luther Stevenson, *Claudius Ptolemy: The "Geography"* (1932; reprint, New York: Dover Publications, 1987), 38.
2. John Rigby Hale, *Renaissance Europe, 1480–1520* (London: Fontana, 1971), 52.

2. Coordinating the World

1. There is an extended discussion of the Hereford map in Paul D. A. Harvey's *Mappa Mundi: The Hereford World Map* (London: Hereford Cathedral and the British Library, 1996).
2. Stevenson, *Claudius Ptolemy*, 39.
3. The zonal *mappaemundi* were circular and divided into five or seven climatic zones. They drew on classical authors such as Ptolemy, and especially on the writings of Macrobius in the early fifth century, as well as on the modifications of later Arab scholars. More than 150 copies are extant. These are more diagrammatic than cartographic and represent an early form of scientific discourse in cartographic depiction. The transitional maps that date from the fourteenth century reflect the more accurate knowledge of the world gained from the maritime charts. They form a bridge to the maps of the Renaissance and mark a shift from a religious to a more humanist iconography. For more on medieval world maps, see Evelyn Edson, *Mapping Time and Space: How Medieval Mapmakers Viewed Their World* (London: British Library, 1997); and David Woodward, "Reality, Symbolism, Time, and Space in Medieval World Maps," *Annals of Association of American Geographers* 75 (1985): 510–21.
4. The wealthy collector Huntington wrote to his agents in London in a letter dated July 15, 1924, asking them to buy twenty items coming up for auction. Next to the Wilton Codex he wrote, "I especially want this." He paid £3,000 for it, a hefty sum at the time.
5. Stevenson, *Claudius Ptolemy*, 38.

3. Encompassing the World

1. Apian's work is placed in a wider context in an online exhibition from the Museum of the History of Science at the University of Oxford: http://www.mhs.ox.ac.uk/measurer/text/contents.htm

2. Little is known about Pomponius Mela, a Roman from southern Spain. His work *De chorographia,* reputedly written around 50 C.E., was largely a description of the known world. He assumed a spherical earth, which was divided into three continents and five climatic zones. In 1478 this work was published in Latin, without diagrams or maps, as a survey of Europe.

The work of Proclus (ca. 410–85) on the sphere and calculations of latitude and longitude was translated into several editions. In 1573 in Florence, Egnazio Danti published an Italian translation as *Sphaera.* Proclus's work was also annotated by Marsilio Ficino.

Sacrobocso (1100?–1256?), also known as John of Holywood, wrote *De sphaera* (The sphere) around 1220. Based on the work of Ptolemy, it begins with the definition of a sphere, discusses the four elements, and outlines a basic cosmography. This book was published many times with many different commentaries in the late fifteenth century and throughout the sixteenth century. Oronce Fine published a copy.

3. Record was born in Wales around 1510. We know little about his early life but we do know that he graduated in 1531 from Oxford with a B.A. and received his M.D. from Cambridge in 1545. By 1547, he was practicing medicine in London.

He was put in charge of the Bristol Mint and came into the web of court intrigue. When he refused to divert money to one court faction, he was accused of treason by Sir William Herbert, later Earl of Pembroke, and was confined to court for sixty days. From 1551 to 1553, he was surveyor of mines in Ireland and supervisor of the Dublin Mint, but he was unsuccessful and lost his job. In 1556 he attempted to regain his position at court and laid charges against Pembroke, who in turn sued for libel in 1556. Record was fined £1,000 and imprisoned for inability to pay the debt. He died in Fleet prison in 1558.

Record was more successful as a Tudor mathematician and was concerned with using mathematical knowledge to solve everyday problems. He published the first English book of arithmetic, *Grounde of Artes,* in 1542. The book was printed periodically until 1699. He also published the first English book on geometry, *Pathway to Knowledge,* in 1551 and is the inventor of the = sign for "equal." A firm believer in practical knowledge, he gave public lectures in London, sponsored by the Muscovy Company, to train the first generation of English long-distance sea navigators.

4. See Hans Wolff, "Martin Waldseemuller: The Most Important Cosmographer in a Period of Dramatic Scientific Change," in *America: Early Maps of the New World,* ed. Hans Wolff (Munich: Prestel, 1992), 111–26.

5. One of his books, *The Messias of the Christians and the Jewes,* was originally

written in Hebrew. It was in the form of a dialogue between an "obstinate Jew" who argues against Jesus being the true Messiah and a Christian who tries to persuade him otherwise. The Jew argues that the words of the prophets are not fulfilled in Jesus, a magician who used "Magick Arts" and broke the Sabbath. The Jew is allowed to put up a good case and gives as good as he gets, so the book cannot be dismissed as a standard piece of Christian anti-Semitic propaganda. It is a strange little piece, more argumentative than religious.

6. I have drawn heavily from the introductory notes in Roger Schlesinger and Arthur P. Stabler, eds., *Andre Thevet's North America: A Sixteenth-Century View* (Kingston: McGill-Queen's Univ. Press, 1986).

7. Thevet's work thus brings into sharp focus the fact that all cosmographies are part fiction: stories with an agenda, narratives with a purpose. Narratives are not so much disproven as they are replaced and reworked. Frank Lestringant describes Thevet's work as "fictonalizing geography." See Frank Lestringant, *Mapping the Renaissance World: The Geographical Imagination in the Age of Discovery* (Berkeley: Univ. of California Press, 1994).

8. For a range of interpretations see Peter J. French, *John Dee: The World of an Elizabethan Magus* (London: Routledge & Kegan Paul, 1972); Richard Deacon, *John Dee: Scientist, Geographer, Astrologer, and Secret Agent to Elizabeth I* (London: Frederick Muller, 1968); and Nicholas H. Clulee, *John Dee's Natural Philosophy: Between Science and Religion* (London: Routledge, 1988); Eva Germaine Rimington Taylor, *Tudor Geography, 1485–1583* (London: Methuen, 1930). A more recent publication that fits Dee's work into a wider context is William H. Sherman's *John Dee: The Politics of Reading and Writing in the English Renaissance* (Amherst: Univ. of Massachusetts Press, 1995).

9. Dee's life can be seen as a three-act tragedy. In the first act, from 1551 to 1582, he is one of the leaders of the English intellectual Renaissance—well-connected, well-read, and influential. He receives royal patronage from Edward VI, but during the reign of the Catholic monarch Mary Tudor (1553–58) he is imprisoned for treachery and conjuring. Dee argues his own case and goes free. The accession of Elizabeth in 1558 marks an upward swing in Dee's affairs. By 1570 he has moved to Mortlake near London, where his house becomes a library meeting place for like-minded intellectuals from home and overseas.

In the second period, from 1582 to 1589, Dee becomes connected with Edward Kelley (1555–95), who turns up at Dee's door in 1582. Kelley is a shady figure; his ears had been cropped two years earlier for forging currency. He offers his services as a spirit medium able to speak with angels. Dee and Kelley hold daily seances calling up angels, and in 1583 they travel to the court of Emperor Rudolf II of Bohemia, where they perform angelic conferences and alchemical transmutations. The relationship sours and Dee returns home in 1589.

This return to England marks the beginning of the third period in Dee's life, in which an aging Dee loses his grip on the center of power. His friends and patrons have died or fallen out of favor; overseas expansion is no longer such an important royal project; and the death of Elizabeth and accession of James in 1603 mark a downturn in Dee's relationship with the monarchy. James is obsessed with witchcraft; he fears plots against him based on occult practices and in the first year of his reign orders Parliament to enact statutes to "uproot enchanters." Dee, friendless and powerless, is an easy target. In a petition to the king in 1604, Dee responds to the "slander" that he was a "conjurer or caller or invocator of divels." The Renaissance magi are now feared figures. Dr. Faustus and his diabolic contract become an evocative image. Giovanni Battista Porta is examined by the Inquisition in 1580, 1592, and 1598, and his work is banned. During the third act of Dee's life, the witchcraft craze is at its peak. The great astronomer Kepler has to defend his mother against the charge, and in 1600 in Rome, Giordano Bruno is tried for heresy, excommunicated, and burnt at the stake. A cloud of reaction is closing out freedom of thought. Dee's final days are sad. His daughter has to sell off his books in order to buy food and heating. Dee dies in poverty in 1608.

10. His library included works by Alhazen, Apian, Roger Bacon, Bellaforest, Borrhaus, Camden, Cartier, Columbus, Contarini, Gilbert, Frobisher, Frisius, Mercator, Munster, Fine, Ptolemy, Ramusio, Schoner, Stoeffler, Thevet, and Vespucci.

11. The work of Frances Yates grapples with the occult philosophies of the Renaissance. See her *Giordano Bruno and the Hermetic Tradition* (London: Routledge & Kegan Paul, 1964); *Theatre of the World* (London: Routledge & Kegan Paul, 1969); and *The Occult Philosophy in the Elizabethan Age* (London: Routledge & Kegan Paul, 1979).

12. A sympathetic account is available in Joscelyn Godwin, *Robert Fludd: Hermetic Philosopher and Surveyor of Two Worlds* (Boulder, Colo.: Shambala, 1979).

4. Mapping the World

1. Marcel van den Broecke, Peter van der Krogt, and Peter Meurer, eds., *Abraham Ortelius and the First Atlas* (Utrecht, The Netherlands: HES Publishers, 1998).

2. Ann Blair, *The Theater of Nature: Jean Bodin and Renaissance Science* (Princeton, N.J.: Princeton Univ. Press, 1997). See also Lynda G. Christian, *Theatrum Mundi: The History of an Idea* (New York: Garland, 1987).

3. Later editions grew in size. An English edition, first published as *The Theatre of the Whole World* by John Norton in 1606 and dedicated to King James, includes maps of the South Pacific, New Spain, Cuba, Peru, and Florida.

4. Mercator's *Atlas* became the template for the Dutch grand atlases of the seventeenth century. The zenith was Willem Blaeu's *Atlas Major*. Blaeu (1571–1638) was a mathematician, astronomer, and instrument maker. Around 1605 he established a printing press near the center of Amsterdam, where he published various works on navigation, astronomy, and theology. He worked with his son Joan Blaeu (1596–1673), and together they published a magnificent series of atlases. The first, published in 1630, consisted of sixty plates, some original and some copied. (A year

earlier, Willem bought several copperplates from Jodocus Hondius.) In 1631 they printed an expanded version, and in 1635 they produced a massive atlas consisting of 208 maps in two volumes titled *Novus Atlas*, with an alternative title of *Theatrum*. This atlas was enormously successful. The Blaeus built upon their success and their business grew. Willem Blaeu died in 1638, but his son Joan carried on the business and assumed his father's appointment as official cartographer to the East India Company. The *Novus Atlas* was expanded to three volumes in 1640 and was continually enlarged in successive printings until six volumes were produced in 1655. Between 1663 and 1665 the 600-map *Atlas Major* (sometimes referred to as the *Grand Atlas*) was produced. This marked the magnificent apex of seventeenth-century Dutch cartography. It is still the largest atlas ever produced—not so much a coffee table book or even a coffee table—almost a small coffee shop. The atlas quite literally represented the world to the literate public.

5. Ian C. Cunningham, ed., *The Nation Survey'd: Essays on Late Sixteenth-Century Scotland as Depicted by Timothy Pont* (Edinburgh: Tuckwell Press, 2001).

6. Bouguereau died in 1596; his partner, Jean Le Clerc, took over the project and saw it through seven editions from 1619 to 1632.

7. One of the first engraved maps of the country, and one revealing something of the religious tensions of the era, was done in 1546 by George Lily, a Catholic. He left Oxford without a degree and went to Avignon; he later joined the household of Reginald Pole, the English Catholic cardinal, in Venice around 1535. Lily was outlawed but returned to England after Mary Tudor's accession. While he was in Europe in exile, Lily drew the map that was the basis for the engraving; he probably was working at the Vatican on cartographic projects. The map was engraved in Italy but the plate was brought to England and copies were made. The map includes a key of towns, a scale, and a grid of latitude and longitude. Drawing heavily on the fourteenth-century Gough map, it depicts towns, mountains, two large forests—"Schirwood" (Sherwood) and "Dene" (Forest of Dene)—and also three small Hampshire towns—Alton, Odiham, and Fordinbridge. This was the area of Lily's family lands, which were confiscated in 1543. The map bore the imprint of a nostalgic remembrance of happier times and places. See Edward Lynam, *The Map of the British Isles of 1546* (Jenkintown, N.J.: Tall Tree Press, 1934).

8. Richard Helgerson, *Forms of Nationhood: The Elizabethan Writing of England* (Chicago and London: Univ. of Chicago Press, 1992).

9. See Helgerson, *Forms of Nationhood*, and Claire McEachern, *The Poetics of English Nationhood, 1590–1612* (Cambridge: Cambridge Univ. Press, 1996).

5. Navigating the Seas

1. Helen M. Wallis, ed., *The Maps and Text of the "Boke of Idrography" Presented by Jean Rotz to Henry VIII, Now in the British Library* (Oxford: Oxford Univ. Press for the Roxburghe Club, 1981), 8.

2. Edward Luther Stevenson, *Portolan Charts: Their Origin and Characteristics* (New York: Knickerbocker Press, 1911), 13.

3. Derek Howse and Norman Thrower, eds., *A Buccaneer's Atlas: Basil Ringrose's South Sea Waggoner* (Berkeley: Univ. of California Press, 1992), 22.

4. C. Koeman *The History of Lucas Janszoon Waghenaer and his "Spieghel der Zeevaerdt"* (Lausanne, Switzerland: Sequoia S.A., 1964).

5. David W. Waters, *The Art of Navigation in England in Elizabethan and Early Stuart Times* (New Haven, Conn.: Yale Univ. Press, 1958), 104.

6. Surveying the Land

1. However, the use of the term *survey* to refer to a written record persisted. John Stow's *Survey of London,* completed in 1598, enlarged in 1618, and published in final form in 1622, used the term *survey,* but the book contained no maps, being a list of churches, monuments, towers, and castles.

2. It was not solved in practical terms until John Harrison created his time pieces in the eighteenth century. The story is covered in Dava Sobel's *Longitude: The True Story of a Lone Genius who Solved the Greatest Scientific Problem of His Time* (New York: Walker, 1995).

3. The literature is summarized in David Hackett Fischer, *The Great Wave: Price Revolutions and the Rhythm of History* (Oxford: Oxford Univ. Press, 1996).

4. Leonard Digges took part in Wyatt's rebellion in 1554. Condemned to death, Digges was pardoned through the help of powerful friends. He published a variety of works on meteorology, gunnery, navigation, and surveying. His son, Thomas Digges, was a Copernican astronomer and author of a military book *Stratiotcos,* published in 1579, which detailed gunnery experiments. Thomas Digges was active in public affairs and, as a member of Parliament, was involved in a plan to repair Dover harbor. He was also a well-known astronomer who experimented with optics.

7. Annexing Territories

1. These have been analyzed in detail by Barbara E. Mundy in her book *The Mapping of New Spain: Indigenous Cartography and the Maps of the "Relaciones Geográficas"* (Chicago: Univ. of Chicago Press, 1996).

SELECTED BIBLIOGRAPHY

Adams, Thomas Randolph, and David Watkin Walters. *English Maritime Books Printed Before 1801.* Providence, R.I.: John Carter Brown Library, 1995.

Alpers, Svetlana. *The Art of Describing: Dutch Art in the Seventeenth Century.* Chicago: Univ. of Chicago Press, 1983.

Aston, Margaret, ed. *The Panorama of the Renaissance.* London: Thames and Hudson, 1996.

Bagrow, Leo. *History of Cartography.* 2nd ed. Chicago: Univ. of Chicago Press, 1985.

Barber, Peter, and Christopher Board, eds. *Tales from the Map Room: Fact and Fiction about Maps and Their Makers.* London: BBC Books, 1993.

Bendall, A. Sarah. *Maps, Land, and Society: A History with a Cartobibliography of Cambridgeshire Estate Maps, ca. 1600–1836.* Cambridge: Cambridge Univ. Press, 1992.

Berggren, J. Lennart, and Alexander Jones. *Ptolemy's "Geography": An Annotated Translation of the Theoretical Chapters.* Princeton, N.J.: Princeton Univ. Press, 2000.

Blair, Ann. *The Theater of Nature: Jean Bodin and Renaissance Science.* Princeton, N.J.: Princeton Univ. Press, 1997.

Brotton, Jerry. *Trading Territories: Mapping the Early Modern World.* London: Reaktion Books, 1998.

Brown, Lloyd Arnold. *The Story of Maps.* New York: Dover, 1977.

Bucher, Bernadette. *Icon and Conquest: A Structural Analysis of the Illustrations of de Bry's Great Voyages.* Chicago: Univ. of Chicago Press, 1981.

Buisseret, David, ed. *Monarchs, Ministers, and Maps: The Emergence of Cartography as a Tool of Government in Early Modern Europe.* Chicago: Univ. of Chicago Press, 1992.

———, ed. *Rural Images: Estate Maps in the Old and New Worlds.* Chicago: Univ. of Chicago Press, 1996.

———, ed. *Envisioning the City: Six Studies in Urban Cartography.* Chicago: Univ. of Chicago Press, 1998.

Campbell, Tony. *The Earliest Printed Maps, 1472–1500.* London: The British Library, 1987.

Canny, Nicholas, ed. *The Origins of Empire: British Overseas Enterprise to the Close of the Seventeenth Century,* vol. 1 of *The Oxford History of the British Empire.* Oxford: Oxford Univ. Press, 1998.

Chamberlain, Paul G. "The Shakespearian Globe: Geometry, Optics, Spectacle." *Environment and Planning D-Society and Space* 19:3 (June 2001): 317–33.

Christian, Lynda G. *Theatrum Mundi: The History of an Idea.* New York: Garland, 1987.

Clulee, Nicholas H. *John Dee's Natural Philosophy: Between Science and Religion.* London: Routledge & Kegan Paul, 1988.

Conley, Tom. *The Self-Made Map: Cartographic Writing in Early Modern France.* Minneapolis, Minn.: Univ. of Minnesota Press, 1996.

Cormack, Lesley B. *Charting an Empire: Geography at the English Universities, 1580–1620.* Chicago: Univ. of Chicago Press, 1997.

Cosgrove, Denis E. *Apollo's Eye: A Cartographic Genealogy of the Earth in the Western Imagination.* Baltimore, Md.: Johns Hopkins Univ. Press, 2001.

Crosby, Alfred W. *The Measure of Reality: Quantification and Western Society, 1250–1600.* Cambridge: Cambridge Univ. Press, 1997.

Cunningham, Ian C., ed. *The Nation Survey'd: Essays on Late Sixteenth-Century Scotland as Depicted by Timothy Pont.* Edinburgh: Tuckwell Press, 2001.

Deacon, Richard. *John Dee: Scientist, Geographer, Astrologer, and Secret Agent to Elizabeth I.* London: Frederick Muller, 1968.

Dekker, Elly, and Peter van der Krogt. *Globes from the Western World.* London: Zwemmer, 1993.

Delano-Smith, Catherine, and Roger J. P. Kain. *English Maps: A History.* London: The British Library, 1999.

Eamon, William. *Science and the Secrets of Nature.* Princeton, N.J., 1991.

Edgerton, Samuel Y. *The Heritage of Giotto's Geometry: Art and Science on the Eve of the Scientific Revolution.* Ithaca, N.Y.: Cornell Univ. Press, 1991.

Edson, Evelyn. *Mapping Time and Space: How Medieval Mapmakers Viewed Their World.* London: British Library, 1997.

Elliot, James. *The City in Maps: Urban Mapping to 1900.* London: The British Library, 1987.

Evans, Ifor M., and Heather Lawrence, eds. *Christopher Saxton: Elizabethan Map-Maker.* Wakefield, England: Wakefield Historical Publications, 1979.

French, Peter J. *John Dee: The World of an Elizabethan Magus.* London: Routledge & Kegan Paul, 1972.

Fischer, David Hackett. *The Great Wave: Price Revolutions and the Rhythm of History.* Oxford: Oxford Univ. Press, 1996.

Godwin, Joscelyn. *Robert Fludd: Hermetic Philosopher and Surveyor of Two Worlds.* Boulder, Colo.: Shambala, 1979.

Goss, John. *Blaeu's "The Grand Atlas of the 17th Century World."* New York: Rizzoli, 1991.

Grafton, Anthony, with April Shelford and Nancy Siraisi. *New Worlds, Ancient Texts: The Power of Tradition and the Shock of Discovery.* Cambridge, Mass: Belknap Press, 1992.

Harley, John Brian. *Maps and the Columbian Encounter.* Milwaukee, Wis.: Univ. of Wisconsin Press, 1990.

———. *The New Nature of Maps.* Baltimore, Md.: John Hopkins Univ. Press, 2001.

Harley, John Brian, and David Woodward, eds. *History of Cartography.* Chicago: Univ. of Chicago Press, 1987.

Harvey, Paul D. A. *The History of Topographical Maps.* London: Thames and Hudson, 1980.

———. *Medieval Maps.* London: The British Library, 1991.

———. *Maps in Tudor England.* London: The British Library, 1993.

———. *Mappa Mundi: The Hereford World Map.* London: Hereford Cathedral and the British Library, 1996.

Helgerson, Richard. *Forms of Nationhood: The Elizabethan Writing of England.* Chicago: Univ. of Chicago Press, 1992.

Howse, Derek, and Norman Thrower, eds. *A Buccaneer's Atlas: Basil Ringrose's South Sea Waggoner.* Berkeley: Univ. of California Press, 1992.

Jardine, Lisa. *Worldly Goods: A New History of the Renaissance.* London: Macmillan, 1996.

Johnson, Hildegard Binder. *Carta Marina: World Geography in Strassburg, 1525.* Minneapolis, Minn.: Univ. of Minnesota Press, 1963.

Kain, Roger J. P., and Richard R. Oliver. *The Cadastral Map in the Service of the State.* Chicago: Univ. of Chicago Press, 1992.

Karrow, Robert W., Jr. *Mapmakers of the Sixteenth Century and Their Maps.* Chicago: Speculum Orbis Press, 1993.

Kitchen, Frank. "John Norden (ca. 1547–1625): Estate Surveyor, Topographer, County Mapmaker, and Devotional Writer." *Imago Mundi* 49 (1997): 43–61.

Koeman, Cornelis. *The History of Abraham Ortelius and His "Theatrum Orbis Terrarum."* Lausanne, Switzerland: Sequoia S.A., 1964.

———. *The History of Lucas Janszoon Waghenaer and His "Spieghel der Zeevaerdt."* Lausanne, Switzerland: Sequoia S.A., 1964.

Lamb, Ursula. *The Cosmographies of Pedro de Medina.* Madrid: Editorial Castalia, 1966.

———. *The Globe Encircled and the World Revealed.* Aldershot, England: Variorum, 1995.

Lenoir, Timothy, ed. *Inscribing Science.* Palo Alto, Calif.: Stanford Univ. Press, 1998.

Lestringant, Frank. *Mapping the World: The Geographical Imagination in the Age of Discovery.* Berkeley: Univ. of California Press, 1994.

Lynam, Edward. *The Map of the British Isles of 1546.* Jenkintown, N.J.: Tall Tree Press, 1934.

Mangani, Giorgio. "Abraham Ortelius and the Hermetic Meaning of the Cordiform Projection." *Imago Mundi* 50 (1998): 59–82.

Masters, Roger D. *Fortune is a River: Leonardo Da Vinci and Niccolo Machiavelli's Magnificent Dream to Change the Course of Florentine History.* New York: Free Press, 1998.

McEachern, Claire. *The Poetics of English Nationhood, 1590–1612.* Cambridge: Cambridge Univ. Press, 1996.

Miller, Naomi. "Mapping the City: Ptolemy's 'Geography' in the Renaissance." In *Envisioning the City: Six Studies in Urban Cartography*, edited by D. Buisseret, 34–73. Chicago: Univ. of Chicago Press, 1998.

Morgan, Victor. "The Cartographic Image of the Country in Early Modern England." *Transactions of Royal Historical Society*, 5th ser., 29 (1979): 129–54.

Mundy, Barbara E. *The Mapping of New Spain: Indigenous Cartography and the Maps of the "Relaciones Geográficas."* Chicago: Univ. of Chicago Press, 1996.

Nicholl, Charles. *The Chemical Theatre.* London: Routledge & Kegan Paul, 1980.

———. *The Creature in the Map: A Journey to El Dorado.* London: Jonathan Cape, 1995.

Nicholson, Nigel. *The Counties of Britain: A Tudor Atlas by John Speed.* London: Thames and Hudson, 1988.

Parker, John. *Richard Eden, Advocate of Empire.* Minneapolis, Minn.: Univ. of Minnesota Press, 1991.

Parry, John Horace. *The Age of Reconnaissance: Discovery, Exploration, and Settlement, 1450–1650.* London: Weidenfeld and Nicholson, 1963.

Quinn, David B. *Set Fair for Roanoke: Voyages and Colonies, 1584–1606.* Chapel Hill, N.C.: Univ. of North Carolina Press, 1985.

Richeson, Allie Wilson. *English Land Measuring to 1800: Instruments and Practices.* Cambridge: Society for the History of Technology and M.I.T. Press, 1966.

Russell, Peter Edward. *Prince Henry "The Navigator": A Life.* New Haven, Conn., 2000.

Schlesinger, Roger, and Arthur P. Stabler, eds. *Ande Thevet's North America: A Sixteenth-Century View.* Kingston and Montreal: McGill-Queen's Univ. Press, 1986.

Schulz, J. "Jacopo de Barbaris's View of Venice: Mapmaking, City Views, and Moralized Geography Before the Year 1500." *Art Bulletin* 60 (1978): 425–74.

Sherman, William H. *John Dee: The Politics of Reading and Writing in the English Renaissance.* Amherst: Univ. of Massachusetts Press, 1995.

Shirley, Rodney W. *The Mapping of the World: Early Printed World Maps, 1472–1700.* London: Holland Press, 1983.

Short, John Rennie. "Maps and the Renaissance." *Journal of Historical Geography* 24 (1998): 360–63.

———. *Alternative Geographies.* Harlow, England: Prentice Hall, 2000.

———. *Representing the Republic.* London: Reaktion, 2001.

Sobel, Dava. *Longitude: The True Story of a Lone Genius Who Solved the Greatest Scientific Problem of His Time.* New York: Walker, 1995.

Staiger, Ralph C. *Thomas Harriot.* New York: Clarion Books, 1998.

Stevens, Henry Newton. *Ptolemy's "Geography": A Brief Account of All the Printed Editions Down to 1730.* London: H. Stevens, Son and Stiles, 1908.

Stevenson, Edward Luther. *Portolan Charts: Their Origin and Characteristics.* New York: Knickerbocker Press, 1911.

———. *Claudius Ptolemy: The "Geography."* 1932. Reprint, New York: Dover Publications, 1987.

Strauss, Gerald. *Sixteenth-Century Germany: Its Topography and Topographers.* Madison, Wis.: Univ. of Wisconsin Press, 1959.

Taylor, Eva Germaine Rimington. *Tudor Geography, 1485–1583*. London: Methuen, 1930.

———. *The Mathematical Practitioners of Tudor and Stuart England*. Cambridge: Cambridge Univ. Press, 1954.

Turner, Robert. *Elizabethan Magic*. Longmead, England: Element, 1989.

Tyacke, Sarah, ed. *English Map-Making, 1500–1650*. London: The British Library, 1983.

Tyacke, Sarah, and John Huddy, eds. *Christopher Saxton and Tudor Map-Making*. London: The British Library Reference Division, 1980.

van den Broecke, Marcel, Peter van der Krogt, and Peter Meurer, eds. *Abraham Ortelius and the First Atlas*. Utrecht, The Netherlands: HES Publishers, 1998.

Vos, Alvin, ed. *Place and Displacement in the Renaissance*. Binghamton, N.Y.: Medieval and Renaissance Texts and Studies, 1995.

Wallis, Helen M., ed. *The Maps and Text of the "Boke of Idrography" Presented by Jean Rotz to Henry VIII, now in the British Library*. Oxford: Oxford Univ. Press for the Roxburghe Club, 1981.

Waters, David W. *The Art of Navigation in England in Elizabethan and Early Stuart Times*. New Haven, Conn.: Yale Univ. Press, 1958.

Wigal, Donald. *Historic Maritime Charts, 1290–1699*. New York: Parkstone Press, 2000.

Williams, J. E. D. *Sails to Satellites: The Origin and Development of Navigational Science*. Oxford: Oxford Univ. Press, 1992.

Wilson, Adrian. *The Making of the Nuremberg Chronicle*. Amsterdam: Nico Israel, 1977.

Wolff, Hans. "Martin Waldseemuller: The Most Important Cosmographer in a Period of Dramatic Scientific Change." In *America: Early Maps of the New World*, edited by H. Wolff. Munich: Prestel, 1992.

Wolter, John Amadelo, and Ronald E. Grim, eds. *Images of the World: The Atlas Through History*. Washington, D.C.: Library of Congress, 1997.

Woodward, David. *Five Centuries of Map Printing*. Chicago: Univ. of Chicago Press, 1975.

———. "Reality, Symbolism, Time, and Space in Medieval World Maps." *Annals of Association of American Geographers* 75 (1985): 510–21.

———, ed. *Art and Cartography*. Chicago: Univ. of Chicago Press, 1987.

Yates, Frances A. *Giordano Bruno and the Hermetic Tradition*. London: Routledge & Kegan Paul, 1964.

———. *Theatre of the World*. London: Routledge & Kegan Paul, 1969.

———. *The Occult Philosophy in the Elizabethan Age.* London: Routledge & Kegan Paul, 1979.

Zandvliet, Kees. *Mapping for Money: Maps, Plans, and Topographic Paintings and Their Role in Dutch Overseas Expansion During the 16th and 17th Centuries.* Amsterdam: Batavian Lion International, 1998.

INDEX

Italic page number denotes illustration.

absolutism, 92, 100
Abu Ma'shar, 62
Admiranda narratio (de Bry), *144, 145*
Aezler, Jacob, 25
Agas, Ralph, 135–36
Agnese atlases, *114*
Agnese, Batista, 114
Aja ib al Makhlaqat (al Qazwini), 48
Alberti, Leon Battista, 68, 70, 101–2
alchemy, 7, 40, 62, 65, 155
Alexandria, Egypt, 11–12
Algiers, *108*
Almagest (Ptolemy), 13, 16, 20, 28
Alternative Geographies (Short), 6
Ambassadors, The (Holbein), 28
America, 33, 47, 50–51, 56, 57, 74. *See also* New World; North America
Antoniszoon, Cornelis, 104
Apian, Peter: approach of, 4, 34–36; astrological work of, 62; influence of, 45, 113, 133; instruction on globes, 30; students of, 27; technical gaze of, 37–42; use of Waldseemuller's maps, 51; works of, 44
Arab world, 15–17, 48–49
armillary sphere, *38*, 38–39

art: cartography and, 9–13, 68–71, *69*, 101–4; estate maps as, 133; exploration and, 143, 144–46; hydrography and, 115. *See also* illustrations
Arte de Navegar (Cortes), 122, *123*, 124–25
Arte de Navegar (Medina), 6, 122, *124*
Ashley, Anthony, 119
astrology: Dee's involvement with, 65; definition of, 48; impact on geography, 7; influence on cosmography, 34–35, 41, 42, 44, 62–63, 155; Ptolemy's contribution to, 13; in Record's work, 47
Astronomicum Caesareum (Apian), 41
astronomy, 41, 48, 151, 153, 155
As You Like It (Shakespeare), 74, 154
Atlas Major (Blaeu), 164–65n. 4
Atlas Minor (Hondius), 84
Atlas of England and Wales (1579, Saxton), 87–91, *89, 90,* 154
Atlas of Europe (Mercator), 80
Atlas sive cosmographicae meditationes de fabrica mundi et fabricati figura (Mercator), 77–78, *81, 82*

atlas(es): of America, 33; Blaeu's, 164–65n. 4; of cities, 105–8; coining of term, 80–81; early printings of, 71; emergence of, 5; as feature of Renaissance world, 2; 1513 edition of *Geography*, 51; *Geography* as, 21; maritime, 116–20; Mercator's, 77–84, 164n. 4; of nations, 84–101; Ortelius's, 73–77, 163n. 3; of portolan charts, 110, 111–16; of world, 72–84

Australia, maps of, 112–13

Bacon, Roger, 62
Barbari, Jacopo de', 103–4
Barentz, Willem, 120
al-Battani, Abu Abdullah, 16
Behaim, Martin, 28
Bellaforest, Francoise de, 54, 58
Belli, Silvio, 6, 130–31
Benese, Richard, 135
Bienewitz, Peter. *See* Apian, Peter
bird's-eye-view maps, 101, 102–4, 105, 106, 107
Blaeu, Joan, 164–65n. 4
Blaeu, Willem, 120, 164n. 4
Blundeville, Thomas, 35–36, 45, 47–48, 62, 63, 126
Boke of Idrography (Rotz), 109–10, 113–14
Boke of Surveying, The (Fitzherbert), 135
books. *See* atlases; writings; *specific book title*
Borrhaus, Martin, 44
Bouguereau, Maurice, 85, 165n. 6
Bourne, William, 125
Braun, Georg, 5, 105–7
Briefe and True Report of the New Found Land of Virginia, A (Harriot), 143
Briefe Collection and compendius extract of strange and memorable things gathered out of the Cosmography of Sebastian Munster, A, 54
Briefe description of universal mapps and cardes and their use, A (Blundeville), 126
Britannia (Camden), 91, 92, 137
Brouscon, Guillaume, 112
Bry, Theodor de, 67, 144
Bucknick, Arnold, 23
Burghley, Lord (William Cecil), 87–88, 91, 94, 95, 125, 135

Cabot, Sebastian, 66
Calderini, Domizio, 23
Camden, William, 91, 92, 96, 137
Canons et documens tresamples, Les (Fine), 62
Cantino chart, 111
Cardinal Bandinello Sauli, His Secretary, and Two Geographers (del Piombo), 1–2, *2*, 20, 156
Carta Marina (Waldseemuller), 51
cartographers, 5. *See also specific cartographer*
cartographic literacy, 6–7, 8, 68, 99, 151
cartographic revolution, 99
cartography, 6, 68–71, 71, 101–4, 133, 142–49
Castle of Knowledge, The (Record), 31, 45–47, *46*, 48
Cecil, William. *See* Burghley, Lord (William Cecil)
Certaine Errors in Navigation (Wright), 126–27, *127*
Charles V (emperor of Holy Roman Empire), 41
Cheney estate map, 136
chorography, 34, 48, 92–93
Chronographia, De (Mela), 162n. 2
Chrysoloras, Emanuel, 20–21

cities, 101, 105, 107–8
city maps, 70, 97–98, 101–8, *103, 106, 137*, 147
Civitates Orbis Terrarum (Braun/Hogenberg), 5, 105–7, *106, 108*, 154
Clavus, Claudius, 21
climatic zones diagram, *44*
colonialism: cartography and, 3, 6, 142–49; charting of seas and, 109, 120; Dee's work and, 36; globes and, 154; impact on geography, 6–7; influence on cosmography, 47, 48, 153; mapping the oceans for, 122–28; as seen through city maps, 108; spatial practices enabling, 7; surveying and, 142–44
Columbus, Christopher, 24, 33, 50, 51
commercial interests: in age of information, 156; charting of seas and, 109–10, 120; envisioning of world and, 151; globes and, 28–30, 154; influence on cosmography, 48, 153; mapping and, 8, 72, 101–8; Norden on, 138–41; as seen through city maps, 108; surveying and, 131–32
commodification of land, 7, 8, 131–33, 138, 141, 153
compass, 19, 61, 110
Compendio del Arte de Navegar (Zamorano), 122
Compleat Surveyor, The (Leybourn), *140*
Contarini, Giovanni Matteo, 23, 33
Cool, Jacob, 73
Copland, Robert, 110
Copperplate Map (London), 107
Cortes, Martin, 122
Cosa chart, 111
cosmo-geographers, 4, 48–66. *See also* Munster, Sebastian; Record, Robert; Thevet, Andre

cosmo-geography, 36
cosmographers, 7, 35–36, 59–67, 79, 150–51. *See also* Apian, Peter; Blundeville, Thomas; Cuningham, William; Fine, Oronce
Cosmographia (Apian): definition of disciplines in, 34; gridded figure in, *130*; instruction on globes in, 30; instrumentation in, *38, 40, 41*; publication of, 4; revision of, 129; technical gaze of, 37–42
Cosmographiae Introductio (Apian), 43, 44
Cosmographiae Introductio (Waldseemuller), 50
Cosmographicae aliquot descriptiones (Stoeffler), 44
Cosmographical Glasse, The (Cuningham), 5, 47, 85, 154
cosmographical sphere, 6
Cosmographie de Lavant (Thevet), 56
Cosmographie unverselle d'Andre Thevet, cosmographe du Roy, La (Thevet), 57–59, *58*
cosmography: applied to colonial enterprise, 145; as applied to surveying, 141; approach of, 4; books of, 37; charting of seas and, 109; counter trends to, 66–67; definition of, 34; emergence of, 5; identifiable elements of, 48–59; influences on, 155; Munster and, 54–56; nature of, 34–35, 81, 163n. 7; splintering of, 66, 151, 153
Cosmography (Munster), 53–56, 76, 104–5, 122–23
cosmo-mysticism, 35–37. *See also* Dee, John
Cosmos (Humboldt), 67
Cuningham, William, 5, 35–36, 45, 47, 85
cyberspace, 156

d'Angiolo, Jacopo, 21, 25
Danti, Egnazio, 70
Davis, Sir Humphrey, 66
Dee, John: accomplishments of, 61–63, 65–66; approach of, 5, 36–37, 155; background of, 60–61, 163–64n. 9; Fine and, 42–43; Frisius and, 130; Mercator and, 79; promotion of navigation by, 125; students of, 61, 65, 66, 133, 134, 142
Del compendio de i secreti rationali (Fioravanti), 65
Delle navigationi et viaggi (Ramusio), 58
demarcation and division of the Indies, A (Velasco), 147
Descriptio Urbis Romae (Alberti), 70, 101–2
Description of the Honor of Windsor, A (Norden), *137*, 138
Descritionis Ptolemaicae augmentum (Wytfliet), 33
Deventer, Jacob van, 105, 107
Dieppe atlases, 109–10, 112–14
Digges, Leonard, 61, 66, 133–34, 166n. 4
Digges, Thomas, 134
Discourse of Western Planting (Hakluyt), 128
Discoverie of Sundrie Errours and Faults daily committed by Landemeaters, Ignorant of Arithmetike and Geometrie, A (Worsop), 135
Donne, John, 7
Dourado, Fernao Vaz, 115
Drake, Sir Francis, 61, 111
drama, 74, 154
Drayton, Michael, 92–93
Durer, Albrecht, *30*, 49, 68, *69*

economic context. *See* commercial interests

Eden, Richard, 66, 122–24
Eeste deel der Zeespiegel (Blaeu), 120
Elementale cosmographicum (Borrhaus), 44
Elements (Euclid), 11, 12
Elizabeth I, queen of England, 62–63, 65, 85–86, 87, 88, 99, *100*
Elizabethan Writing of England, The (Helgerson), 91–92
Enchuyser Zeecaert-Boecke (Waghenaer), 120
enclosure movement, 132–33, 138, 141
England: atlases of, 86–101; city maps of, 105, 107; cosmographers of, 45–48; development of navigational works in, 122–28; surveying of, 132–41
Enkhuizen, Holland, 116
epistemological rupture, 153
Eratosthenes, 12, 13
Escorial Atlas, 147
Esquivel, Pedro de, 147
Essex (map), 95
estate maps, 132–41
Euclid, 12, 150
European maps, 21, 42, 71, 83, 85–101. *See also* England; France; Spain
exploration, 29–30, 32–33, 51, 56, 61, 66. *See also* colonialism

Face of the Earth (al-Khwarizmi), 16
Faerie Queen, The (Spenser), 8
Farrisol, Abraham, 49
Fillastre, Guillaume, 21
Fine, Oronce: approach of, 4, 35–36; astrological work of, 62; background of, 42; Dee and, 60, 66; influence on cosmographers, 45; publication of *De sphaera*, 162n. 2; translations of Apian's work, 44; works of, 36, 42–43, *43*

Fioravanti, Leonardo, 65
Fitzherbert, John, 135
Fludd, Robert, 66–67
Forlani, Paolo, 29
France, 42, 85
Frisius, Gemma: codification of surveying, 129–30; first globe of, 28; influence of, 133; as instrument maker, 39; students of, 60, 66, 79, 113; version of Apian's *Cosmographia*, 37
Frobisher, Martin, 61, 143

Gallucci, Giovanni Paolo, 30–31, 44, *45*
Garcie, Pierre, 110
Gastaldi, Giacomo, 23, 26
gazeteers, 27, 137
General and Rare Memorials (Dee), 61, 66
geographic concerns, 48–59
geography: definition of, 34, 48; emergence of, 5, 59, 67, 151, 153; fusion with history, 3, 18–19, 49–50, 94–95; interchangeability with cosmography, 36; position of in Renaissance, 1; Renaissance writings concerning, 5
Geography (Ptolemy): cartographers working with, 23; city maps in, 101; concept of spherical earth in, 27–28; first printing of, 6; influence on exploration, 32–33; influence on Munster, 52; influence on Renaissance art, 70; information within, 13–17; maps in, *22, 24, 25, 26, 27*; on plane surface maps, 31–32; Rome edition, 23–24; significance of, 1, 3–4, 26, 150; Strasbourg edition, 24–25, *25*, 159; translations/copies of, 20–22, 34, 159–60; Ulm edition, 21, 22, 33, 159;

Venice editions, 25–26, *26*, 159; Waldseemuller's contribution to, 51
geography and general description of the Indies, A (Velasco), 147
Geomance, La (La Taille), 63
Geomancie of Maister Christopher Cattan, The (Sparry), 63–64, *64*
geomancy, 63–64
Geometrical Practise, named Pantometria, A (Thomas Digges), *134*, 134–35
geometry. *See* mathematics
Germanus, Donnus Nicolaus, 21
Gheeraerts, Marcus, the Younger, 99
Ghisolfi, Francesco, 115
Gilbert, Humphrey, 61
Giraldi, Giglio Gregorio, 122
globes, 2, 27–31, *30*, 154–55
Gough map, 86
Great Library (Alexandria), 11–12
Great Volume of Famous and Rich Discoveries, The (Dee), 66
Gregory XIII (pope), 70
Greville, Sir Fulke, 96
grid: construction of, 3–4, 5, 9–13, 150–51; in *Cosmographia* (Apian), *130*; early portolans and, 19; early uses of, 6–7, 25; expression of perspective and, 101; extension of, 33; on globes, 29; Norden's use of, 137; Ortelius's use of, 76; use of in art, 68, *69*, 70
Ground of Artes (Record), 61

Hakluyt, Richard, 61, 80, 127–28
Harriot, Thomas, 61, 66, 125, 142–46
Harrison, John, 166n. 2
Hatton, Sir Christopher, 119
Helgerson, Richard, 91–92
Heneage, Sir Thomas, 94

Henry (king of France), 85
Henry the Navigator, 32
Henry VIII (king of England), 85, 132
Hereford Cathedral world map, 9, *10*, 17–18
Hipparchus, 13
historicall and chorographicall description of Hertfordshire, An (Norden), 137
historicall and chorographicall description of Middlesex, An (Norden), 94–95, 137
history, 3, 18–19, 49–50, 76–77, 94–95, 150
History of Great Britaine (Speed), 98
history of trauvyle, The, 123–24
Hoefnagel, Georg, 107
Hogenberg, Frans, 5, 72, 105–7
Holbein, Hans, 28
Hondius, Henry, 84
Hondius, Jodocus, 23, 83–84, 91, 165n. 4
Hondius, Nicolaus, 23
Hooftman, Gilles, 73
Humboldt, Alexander von, 67
Huntington, Henry E., 161n. 4
hydrography, 5, 47, 109–28

Ibn Abi al-Rijal, 62
iconography, 17–18, 85
al-Idrisi, Abu Abdullah Mohammed Ibn Al-Sharif, 16
al-Idrisi world map, *16*
Igeret Orhot Olam (Farrisol), 49
illustrations: in *Book of Idrography*, 109; on city maps, 104–5; first printed book with, 103; function of, 107; in *Nuremberg Chronicle*, 49; on portolan charts, 112, 114, 116. *See also* art
Imola, map of, *102*
imperialism. See colonialism

instrumentation: construction of the grid and, 151; development of, 4, 6, 38–43, 47; development of portolan charts and, 19; of modern times, 156; of surveying/mapping, 152
Instrument of the Suns, The (Munster), 52
Intended Guyde for English Travailers, An (Norden), 96, 97
Isagoge (Apian), 40

Jansson, Henry, 84
Jansson, Jan, 84, 105
Java, La, 112–13

King-Hamy portolan, 111
knowledge, 151, 153
Koeman, Cornelius, 119
Kolb, Anton, 104
Kratzer, Nicolas, 86–87
al-Khwarizma, Abu Ja'far Muhammad ibn Musa, 16

La Taille, Jean de, 63
latitude/longitude: of European cities, *15*; on globes, 29, *35*; on maps in editions of Ptolemy's *Geometry*, *25*, *26*; in Mercator's *Atlas*, 81–82; perception of world through, 150; on projections, 32; Ptolemy's description of, 13–14; as Ptolemy's legacy, 4
Lautensack, Hans, 104
Le Clerc, Jean, 165n. 6
Leigh, Valentine, 135
Libellus (Frisius), 129
Liber cronicarum (Schedel), *50*, 103
Libro del misurar con la vista (Belli), 6, 130–31, *131*

Light of Navigation, The (Blaeu), 120, *121*, 154
Lily, George, 165n. 7
London maps, 107, 137

magic-hermetic tradition, 61–65, 155
Maner of Measurying, The (Benese), 135
map projections: cordiform, 32, 42, 151; explosion in use of, 31–32; in *Geography*, 23–24; Mercator's, 29, 32, 80, 151; in Mercator's *Atlas*, 82; polar (conic), 70; Ptolemy's, *14*, 27
mappaemundi, 17–19, 161n. 3
mapping/mapmaking: art and, 68–70; of cities, 101–8; commercial/national interests and, 8; in Cuningham's work, 47; development of, 3; Frisius's influence on, 129–30; function of, 155; instruments of, *152*; of nations, 84–101; Ptolemy's instructions for, 15; purpose of, 5–6; surveying and, 135, 136
Mariner's Mirrour (Waghenaer), 6, 116–20, 154
Marquis of Mantua, 70
Massaio, Pietro del, 21, 101
mathematics: in Apian's *Cosmographia*, 40; Dee's approach to, 61; of Euclid/Ptolemy, 12–13; of Frisius, 129; grid construction and, 10, 150; in Record's work, 46; of surveying, 135
Mathematics (Euclid), 61
Mattioli (physician of Sienna), 25
M. Blundeville His Exercises (Blundeville), 47–48, 126
Medici, Cosimo de, 70
Medina, Pedro de, 6, 105, 122
Mela, Pomponius, 40, 162n. 2
Mercator, Gerardus: atlas of, 5, 77–78; background of, 79–80; Dee and, 61, 66; globe making of, 28–29; influence of, 72; Ortelius's view of, 74; portolan projections of, 32; views of, 6; work on Ptolemy's *Geography*, 23
Mercator, Gerardus (grandson), 78
Mercator, Johann, 78
Mercator, Michael, 78
Mercator, Rumold, 78
Mercator's *Atlas*, 77–78, *81*, *82*
Merian, Matthaus, 105
Messias of the Christians, The (Munster), 162–63n. 5
mimetic devices: in Mercator's *Atlas*, 82–83; on Ortelius's maps, 76; on portolan charts, 112; in Saxton's atlas, *90*; on Speed's maps, 98–99; use of in atlases, 70–71, *90*
Mirror of Alchimy, The (Bacon), 62
Molyneux, Emery, 30
moste profitable and commendable science, The (Leigh), 135
Mundi Sphaera, De (Fine), 43
Munster, Sebastian: approach of, 5, 36, 51–52, 54–56, 59; background of, 52; town prospects of, 104–5; use of Waldseemuller's maps, 51; work borrowed from, 107; work on Ptolemy's *Geography*, 23; works of, 52–56, 162–63n. 5

nationalism, 8, 84–88, 91–92, 99–101, 151, 153
Natural Magick (Porta), 65
navigation: in Blundeville's works, 48; books of instruction on, 120–28; commercial/national interests and, 8; in Cuningham's work, 47; development of, 3; emergence of, 5–6, 151, 153; mathematical

navigation (*cont.*)
  practices of, 6; poetic references to, 7–8; portolan charts/atlases for, 109–20; prognostication and, 155
New and Improved Description of the Lands of the World, amended and intended for use of Navigators (Mercator), 32
New World: in *Boke of Idrography*, 109; exploration, 33; first atlas of, 33; lack of in Mercator's *Atlas*, 83; maps of, 22, 23, *24*, *25*, 51, 76, *144*, *145*; Ortelius's maps of, 77; portolan charts recording, 111; representation of, 33; surveying of, 142–46; Thevet's incorporation of, 56–57, 59
Norden, John, 5, 92, 94–96, 136–40
North America, 33, 142–46, *144*, *145*. See also New World
northern orientation, 110
Novus Atlas (Blaeu), 165n. 4
Nuñez, Pedro, 61, 66
Nuremberg Chronicle, 49–50, 103

occult discourses: in Apian's *Cosmographia*, 40; Fine's interest in, 42, 44; in Munster's work, 54–55; Renaissance cosmography and, 4, 5, 61–62. *See also* cosmo-mysticism; Dee, John
Ogilby, John, 91
Orontij Fine Delphinatis (Fine), *43*
Ortelius, Abraham, 5, 61, 66, 72–75, 79, 96

Pannartz, Arnold, 23
panopticon, 104
Paris, Matthew, 86
Pedro (prince of Portugal), 32

Peregrinatio in terram sanctam, 102–3
perspective, 68, 70, 101
Philip II (king of Spain), *75*, 105, 146–47
Piombo, Sebastiano del, 1–2, 20, 156
plaine treatise of the first principles of cosmographie, A (Blundeville), 35, 63
Plancius, Petrus, 29, 91
plane surfaces, 31–32
plan view maps, 101, 102, 106, 107
Pole, Reginald, 165n. 7
political interests: charting of seas and, 109–10, 120; Dee's work and, 36; embedded in maps, 71, 98; mapping and, 87–89, 99–101; production of navigational books and, 122–28
Poly-Olbion (Drayton), 92–93, *93*
Pont, Timothy, 85
Porta, Giovanni Battista, 65
Portolan Atlas (Agnese), *114*
Portolan Atlas (Ghisolfi), *115*
portolan charts/atlases, 18–19, *19*, 32, 110–16, *113*, *114*, *115*, 117
Portrait of a Humanist (del Piombo), 1
Portrait of Queen Elizabeth (Gheeraerts), 99, *100*
Preparative to Platting, A (Agas), 135–36
Principal Navigations, Voyages and Discoveries of the English Nation, The (Hakluyt), 80, 128
printing, 21–23, 71, 102–3, 154
Proclus, 40, 162n. 2
prospect view maps, 101, 104, 105, 106, 107
Ptolemy, Claudius: background of, 11–12; city maps of, 101; influence on geography, 20–27, 150; influence on Renaissance art, 70; legacy of, 26–27; sphericity of earth

Index ❦ 183

and, 27–28; works of, 1, 3–4, 6, 12–17. *See also Geography* (Ptolemy)
Ptolemys (rulers of Alexandria), 11
Ptolemy's Geography (Munster), 52, *53*, 54

al Qazwini, Zakariya Ibn Muhammad, 48–49

Radermaker, Jan, 72–73
Raleigh, Sir Walter, 61, 65, 125, 142–43, 146
Rare Memorials pertaining to the Perfect Arte of Navigation (Dee), 125
Rathborne, Aaron, 6, 140–41
Record, Robert, 31, 35–36, 45–47, 162n. 3
Regiment of the Sea, A (Bourne), 125
Relaciones Geográficas, 147–49, *148*
religion: development of surveying and, 132; factionalism in England, 85–86, 88; influence on cartography, 17–18, 88, 165n. 7; influence on cosmography, 34–35, 155; Mercator and, 79, 81; Munster and, 52; position of in Renaissance, 1
Renaissance: art of, 68–71, 101–4, 115, 143, 144–46; cosmography of, 34–35, 53; geographical thought of, 2–8, 9–13, 20–27
Re Nautica, De (Giraldi), 122
Representing the Republic (Short), 6
Reuwich, Erhard, 103
Richard of Holdingham, 9
Ringmann, Matthew, 25
roads, 90–91, 94
Rome, scaled map of (Alberti), 101–2
Rosselli, Francesco, 104
Rotz, Jean, 109–10, 113–14

Routier de la mer, Le (Gracie), 110
Rudd, John, 87
Ruysch, Johan, 23, 24

Sacrobosco, 40, 162n. 2
Sanderson, William, 30
satellite photography, 155–56
Sauli, Cardinal Bandinello, 1–2
Saxton, Christopher, 5, 85, 87–91, *92*, 136
Schoner, Johannes, 28, 29, 33, 42, 51, 62
Schweizer Chronik (Stumpf), 104
Seckford, Thomas, 87, 88, 91
Seld, Jorg, 104
Shakespeare, William, 74, 154
Sharp, Bartolomew, 111
Silva, Nuno da, 111
Singularitez de la France antarctique, Les (Thevet), 56–57, *57*
social issues, 71, 98–99, 131–32, 136–37, 138–41
space, 3–8, 71, 150–51, 154, 155–56
Spain, 105
Sparry, Francis, 63–64
spatial discourses, 3–8. *See also* colonialism; cosmography; grid; mapping/mapmaking; navigation; surveying
spatial envisioning, 8, 9–13, 26–27, 154
spatiality, 6–7, 29, 155–56
spatial practices, 6–8, 132, 135, 151–53
spatial revolution, 9–13
spatial sensitivity, 7–8, 154
spatial surveillance, 7, 8, 87–89, 104, 105, 129
Speculum Britanniae (unpub., Norden), 94, 154
Speed, John, 5, 96–99, 105–6, 137
Spenser, Edmund, 8
Sphaera, De (Sacrobosco), 162n. 2

sphere, 34, 44, 151
sphericity of earth, 27–31, 34, *35*
Spieghel der Zeevaerdt (Waghenaer), *118, 119*
Stoeffler, Johannes, 44
Stratiotcos (Digges), 166n. 4
Stumpf, Johannes, 104, 107
Survey of London (Stow), 166n. 1
surveying: codification of, 129–32; commercial/national interests and, 8; in Cuningham's work, 47; definition of, 129, 166n. 1; development of, 3, 132–41; emergence of, 6, 151, 153; of England, 87–89, 94; instruments of, *152*; mapmaking and, 136; mathematical practices of, 6; social issues and, 131–32
Surveyor, The (Rathborne), 6, *139, 140–41*
Surveyors Dialogue (Norden), 138–40
Sweynheym, Conrad, 23
Sylvanus, Bernardus, 23

Tavernier, Gabriel, 85
technical gaze, 4, 35, 37–48
Tectonicon (Leonard Digges), 134
Terra Java, 112–13
Tetrabiblos (Ptolemy), 20, 62
theater, 74, 154
Theatre Francois, Le (Bouguereau), 85, *86*
Theatre of the British Empire, The (Speed), 91
Theatre of the Empire of Great Britaine, The (Speed), 96–99, *98*, 105–6, 137, 154
Theatrum Europaeum (Merian), 105
Theatrum Mundi (Gallucci), 30–31, 44, *45*, 154
Theatrum Orbis Terrarum (Ortelius), 5, 72–77, *73, 75, 77, 78*, 106, 154, 164n. 3
Thevet, Andre: approach of, 5, 36, 51–52, 58–59, 163n. 7; background of, 56; geography of, 36; works of, 56–59
Thresoor der Zeevaerdt (Waghenaer), 120
time pieces, 129, 166n. 2
T-O maps. *See* tripartite maps
topography, 81, 153
Traite de la construction et des principaux usages des instruments de mathematique (Bion), *152*
transitional maps, 161n. 3
triangulation. *See* mathematics
trigonometry. *See* mathematics
tripartite maps, 17, *18*
Typus Orbis Universalis (Apian), 40

Uebelin, Georg, 25

"Valediction: Forbidding Mourning, A" (Donne), 7
Vallard Atlas of 1547, 112, *113*
Velasco, Juan Lopez de, 147
Venice maps, 103
Vespucci, Amerigo, 50–51
da Vinci, Leonardo, 70, 102
volvelles, 39, *41*, 42, 44, *45*

Waghenaer, Lucas Janszoon, 6, 116–20
Waldseemuller, Martin: background of, 51; globe of, 29; maps of, *24, 25*; translation of Ptolemy's *Geography*, 23, 24–25; works based on, 40; works of, 50
wall decoration, 17, 70, 91
Wentworth, Michael, 136

Werner, Johannes, 32
White, John, 142–43, 144–46
Wilton Codex, 21, 22, 161n. 4
Woensam, Anton, 105
Worsop, Edward, 135
Wotton Underwood estate map, *133*
Wright, Edward, 125, 126–27
writings: cartography and, 7–8, 71, 74, 154; of cosmo-geographers, 36, 47, 50–59; development of geography and, 5; Helgerson on English works, 91; maps within, 17; promoting navigation, 125–26; of Renaissance cosmographers, 34–35; on use of globes, 29
Wyatt's rebellion, 166n. 4
Wyngaerde, Anton van den, 105, 147
Wytfliet, Cornelius, 33

Zamorano, Rodrigo, 122